Zero – Much to Do About Nothing?

Jim E. Riviere

Zero – Much to Do About Nothing?

Springer

Jim E. Riviere
North Carolina State University
Raleigh, NC, USA

ISBN 978-3-031-82997-0 ISBN 978-3-031-82998-7 (eBook)
https://doi.org/10.1007/978-3-031-82998-7

This Springer imprint is published by the registered company Springer Nature Switzerland AG
The registered company address is: Gewerbestrasse 11, 6330 Cham, Switzerland

To mentors, colleagues, and students I have learned so much from.

"A zero itself is nothing, but without a zero you cannot count anything; therefore, a zero is something, yet zero."

Dalai Lama

Preface

Zero and Nothing. They sound like such simple concepts, but have you ever thought about what they actually mean? When a calculation results in zero, one takes a second look. Why can't you divide by zero? When zero appears as a multiplier in a long equation, all other terms, no matter how complex, magically vanish leaving zero as the result. When you read a food label that says zero sugar or zero pesticides or no additives, what does that really mean? When a rocket countdown reaches zero, lift-off occurs! How do you feel when your bank account is empty, or your assets reach zero, in contrast to the feeling of securing a zero-interest no-fee credit card? When asked what your friend is doing, nothing may be their answer but then one often becomes suspicious. What does emptiness connote in philosophy, and does the concept of nothingness have significance to you? This book addresses these questions and probes why zero and nothing often carry such intellectual and emotional baggage.

The book starts with an introduction covering the vast domains where zero is encountered ranging from thermometers to roulette wheels, to its role in mathematics, physics, deep space, biology, and toxicology to practical issues in food safety and doping. The first chapter delves into the concept of nothingness, emptiness, and the void. It traces how these concepts influenced the development of mathematics and the sciences, paying particular attention to the differences in their acceptance in Western *versus* Eastern philosophies and religions. The history of zero is then traced from its relatively late appearance in ancient numbering systems to the modern era, first as a placeholder and then as a true number, paying particular attention as to why both the concepts of zero and nothingness diverged between different cultures. Zero's central role in enabling the development of higher mathematics is discussed as well as to why its use in certain functions is prohibited. The role of both zero and the null-hypothesis in statistics is probed. Physics is one discipline that attempts to define the lower limits of the physical zero related to absolute zero temperature, to determine if a mass is present, or whether there is truly such a thing as empty space. These discussions reinforce the importance of dimensional scale in understanding why zero may have different implications and meaning in the quantum, atomic, microscopic, and macroscopic worlds.

The next chapters delve further into the applied sciences of chemistry, biology, and toxicology, emphasizing why not being able to chemically detect the presence of a substance, such as a poison, is different than assessing the ability of that substance to cause an adverse effect in a biological system. Every week the media reports finding very low levels of "dangerous" chemicals somewhere in the food chain or environment. But do extremely low levels, at the limits of detection, have anything to do with the type of adverse effects associated with the same chemical at much higher and longer-term exposures? Why in chemistry is detection fundamentally different than quantification? These chapters are mildly technical as some introduction to chemistry and mathematics is required to define the contexts in which zero can be applied and understand the process where a zero concentration in analytical chemistry is determined. The need to define zero risk in food and environment safety should be a required twenty-first-century life skill. The book wraps up with discussions ranging from zero tolerance in sports doping, to end-of-life issues that point to zero biological activity, as well as attempting to provide some guidelines where zero can properly be interpreted.

The audience of this book is targeted at those interested in why such simple concepts, such as zero and nothing, have become central to so many disciplines. Also targeted are readers interested in the infusion of philosophical concepts into the hard as well as applied sciences, and how such concepts have direct relevance in how science is interpreted today. The book targets specialists and regulators in national and international agencies to point out that the meaning of zero is context dependent, and values used in one discipline may not simply be applied to another without potential unintended consequences occurring. Many people intently focused on one discipline do not realize that zero obtained from other sources may have unknown assumptions attached. Creating operational definitions of zero for specific applications is crucial. Finally, in the modern age dominated by Artificial Intelligence and Big Data, scraping zeros from the web independent of how they were generated or used is simply wrong. So, although the title of the book is "Zero – Much to do about Nothing," it turns out one can have a lot to say about nothing!

Raleigh, NC, USA

Jim E. Riviere

Acknowledgements

Numerous families, friends, and colleagues have graciously given their time to read and critique various parts of this book. I thank Professors Ronald Baynes, Majid Douraki-Jaberi, Gerald Wyckoff, and Mark Papich for reading the first draft of the entire book; and Professors Xuan Xu for reviewing the mathematics chapter, Jeffrey Comer the physics chapter, and Santosh Aryl for the chemistry chapter. All caught errors and made great suggestions. Any remaining errors are solely the responsibility of the author. Thanks also go to James Yeattes and Hiroko Enomoto for creating the standard curve comparisons in Figure 5-6 and along with Scott Wetzlich reviewing the chemistry chapter; and Nader Zad for compiling the Maximum residue level (MRL) compilations in Figure 9-6 and Table 9-1. Illustrations were either created by the author, open sourced from Wikimedia Commons, or taken from my textbook on Pharmacokinetics. Finally, I wholeheartedly thank my daughter Jessica Rogers for copy editing the entire book, and my spouse and lifetime professional colleague Distinguished Professor Nancy Ann Monteiro-Riviere for debating concepts and proofing numerous chapters at all stages, as well as putting up with my answering "What are you doing dear?" with "Nothing dear."

Contents

About the Author

Jim E. Riviere, D.V.M., Ph.D., is a Distinguished Professor Emeritus from both North Carolina State and Kansas State Universities and an elected member of the National Academy of Medicine. He has authored/edited 22 books and published over 600 scholarly publications. His honors include the APhA Ebert Prize, FDA's Harvey Wiley Medal, the University of North Carolina System's O Max Gardner award as well as Lifetime Achievement Awards in his professional societies. Professor Riviere has served on numerous national and international committees related to toxicology and food safety. He co-founded the Food Animal Residue Avoidance and Depletion (FARAD) program in 1981. Riviere earned his bachelor's and master's degrees from Boston College and graduate, professional, and honorary doctoral degrees from Purdue University. He has three children and currently resides in Raleigh North Carolina with his spouse of 48 years, Distinguished Professor Nancy Monteiro-Riviere.

Abbreviations

α	Probability of a Type I error (Statistics)
A	Atomic mass number
ACGIH	American Conference of Governmental Industrial Hygienists
AD	Anno Domini (Year of the Lord)
ADI	Acceptable daily intake
ADME	Absorption, distribution, metabolism, and elimination
AI	Artificial intelligence
AUC	Area under the curve
ATP	Adenosine triphosphate molecules
ATSDR	Agency for Toxic Substances and Disease Registry
β	Probability of a Type II error (Statistics)
BBB	Blood brain barrier
BC	Before Christ
BCE	Before Common Era
BEI	Biological exposure indices
BMD	Benchmark dose
BQL	Below quantifiable limit
c	Speed of light
CE	Common Era (aka AD)
CLT	Central Limit Theorem
CODEX	Codex Ailmentarius Commission, an international food standard agency
°C	Degrees Centigrade
°F	Degrees Fahrenheit
°K	Degrees Kelvin
D	Diffusion coefficient
dB	Decibel
DDT	Dichloro-diphenyl-trichloroethane
DES	Diethylstilbesterol
DNA	Deoxyribonucleic acid
D-R	Dose-response
e	Euler's number (mathematics)

ECG	Electrocardiogram
EEG	Electroencephalogram
ELISA	Enzyme-linked immunosorbent assay
EPA	Environmental Protection Agency
FAO	Food and Agricultural Organization
FARAD	Food Animal Residue Avoidance and Depletion program
FDA	Food and Drug Administration
FDCA	Food, Drug and Cosmetic Act
FEI	Federation Equestre Internationale
FSIS	USDA Food Safety and Inspection Service
GLP	Good Laboratory Practices
GC	Gas chromatography
GHS	Globally Harmonized System
GMT	Greenwich Mean Time
H	Membrane thickness
H_0	Null hypothesis (Statistics)
H_1	Alternate hypothesis (Statistics)
HISA	Horseracing Integrity and Safety Authority
HIWU	Horseracing Integrity Welfare Unit
HPLC	High performance liquid chromatography
HPV	Human papillomavirus
IARC	International Agency for Research on Cancer
ID	Intradermal
IDL	Instrument detection limit
IM	Intramuscular
IN	Inhalation
IU	International units
IUPAC	International Union of Pure and Applied Chemistry
IV	Intravenous
K_{ow}	Octanol/water partition coefficient
K_p	Permeability coefficient
LC	Liquid chromatography
LD_{50}	Median lethal dose
ln	Natural logarithm
LOAEL	Lowest observed adverse effect level
LOD	Limit of detection
log	Base 10 logarithm
LOQ	Limit of quantification
LLOQ	Lower limit of quantification
ℓP	Planck length
M	Molarity
MID	Minimal infective dose
ML	Machine language
MLA	Minimum level of applicability
MOA	Mode of action

MRL	Maximum residue level
MS	Mass spectrometry
MTD	Maximum tolerated dose
MTL	Maximum tolerable (or toxic) level
MWQ	Minimum weighable quantity
N	Instrument noise
NASEM	National Academies of Science Engineering and Medicine
NIEHS	National Institute of Environmental Health Sciences
NIOSH	National Institute of Occupational Safety and Health
NOAEL	No observed adverse effect level
NOEL	No observed effect level
NOHSC	National Occupational Health and Safety Commission
NRC	National Research Council
NRP	National Residue Program
NTP	National Toxicology Program
OSHA	Occupational Safety and Health Administration
P	Partition coefficient (lipid/water)
PAH	Polyaromatic hydrocarbons
PCBs	Polychlorinated biphenyls
PCR	Polymerase chain reaction
PEL	Permissible exposure limit
PO	Oral drug administration—per os
PPB	Parts per billion (microgram per kilogram/liter)
PPM	Parts per million (milligram per kilogram/liter)
PPQ	Parts per quadrillion (picogram per kilogram/liter)
PPT	Parts per trillion (nanogram per kilogram/liter)
QSAR	Quantitative structure activity relationship
R	Correlation coefficient
R^2	Coefficient of determination
RfC	Reference concentration
RfD	Reference dose
RSD	Relative standard deviation
RNA	Ribonucleic acid
S	Instrument signal
S/N	Signal-to-noise ratio
SC	Safe concentration
SC	Subcutaneous
SD	Standard deviation
SDS	Safety data sheet
SEM	Standard error of the mean
SOP	Standard operating procedure
TCDD	Tetrachlorodibenzo-p-dioxin
TDMI	Total maximum daily intake
TLM	Tolerance limit method
TOP	Topical

T_P	Planck time
TLV	Threshold limit value
TTC	Threshold of toxicological concern
UPLC	Ultra performance liquid chromatography
USDA	United States Department of Agriculture
USP	United States Pharmacopeia
UT	Universal Time
WADA	World Anti-Doping Agency
WHO	World Health Organization
WOE	Weight-of-evidence
Z	Atomic number
#0dte	Zero days till expiration or a zero-day stock option

List of Figures

List of Tables

Chapter 1
Introduction and Overview

> In the history of culture, the discovery of zero will always stand out as one of the greatest single achievements of the human race. (Tobias Dantzig)

I have always been fascinated by the concept of and definition of *zero*. The idea of *nothing,* as well as its assignment to the specific number 0, is encountered in many fields ranging from mathematics to philosophy, and the pure sciences of physics and chemistry to its practical application in areas including toxicology, food safety, statistics, marketing, and the law. We are constantly exposed to zero in areas far beyond mathematics where we are tempted by zero calorie soda to credit cards with 0% interest. Other counting numbers do not share this unique aura. The number one is widely used, and anything greater than one just often becomes plural. The non-integer numbers π and ∞ come close, but as we will see throughout this book, zero easily assumes prominence in wide ranging swaths of human endeavor (Fig. 1.1).

Fig. 1.1 Zero everywhere

It seems like such a simple question; what does zero mean? Why would anyone care if people can't agree on a definition? Zero is well defined in the foundations of mathematics, but few who use it in their work have access to this. Zero is a misfit amongst the normal so-called counting numbers. It defies definition as it reflects nothing, emptiness, or a void. As will be seen, ancient Greeks abhorred the concept of emptiness and the void. Zero causes headaches when first encountered in arithmetic where one is forbidden to divide by it. Zero occupies a unique position in our

---

Figure Use Attributions
Figs. 1.1, 1.3, 1.4, 1.5, and 1.6: Wikimedia – No restrictions
All other figures are either self-drawn, photographed or from author's own publications (attributed)

J. E. Riviere, *Zero – Much to Do About Nothing?*,
https://doi.org/10.1007/978-3-031-82998-7_1

numbering system as it is the only counting integer that does not relate to any actual quantity. It came into existence centuries after all the counting numbers were introduced in civilizations around the world.

Although one can endlessly debate the impact and significance of a number relating to nothing, problems arise when one must measure zero and attest that nothing is there. A simple example from the above figure illustrates this concept. Zero sugar cola is easily defined as sugar simply not being used in the soda's manufacture. However, if a natural beverage is to be labeled as such, there must be proof that analysis would yield a result of zero. But how precise does that need to be, what type of sugar is being assayed and how much variability is there in natural beverages? This zero will be the focus of the later chapters of this book.

Zero was originally derived from the Arabic word *sifr,* having the same route as *cipher,* meaning nothing. Zero divides the number line between negative and positive numbers (Fig. 1.2).

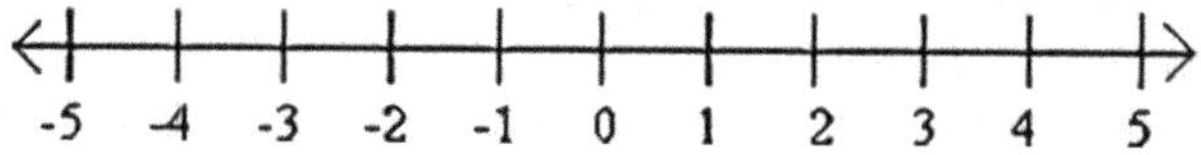

Fig. 1.2 Number line

In arithmetic, zero is the name of the digit "0", also referred to as naught or a cipher. Yet it possesses a unique mystique in that something having a value of zero is nothing or empty; we served all the people in a line and no one remains, we ate all the apples and no apples are left in the basket, we spent all the money and now there is no money in the bank, or we removed all contaminants and there are no chemicals left to harm you! For many facets of everyday life, when simple counting of objects is possible, there is no debate. The definition of zero is simply as such, "we used them all up and nothing is left", or to begin with, "there is nothing present to count or a container is empty". It's when there are objects or entities that *can't be seen but must be indirectly measured by an instrument,* that issues and confusion arise.

Robert Kaplan, a professor of mathematics at Harvard and author of the book *The Nothing That Is: A Natural History of Zero* wrote: "*Zero is in the mind, but not in the sensory world.*" Once zero came along, we had negative numbers. Multiplication became easier. Zero helps us to understand that we use math to think about things that have no counterpart in our physical daily living experience. Some studies have shown that adults take a few moments longer to recognize the number zero compared to other numerals. Zero does hold a unique place in our minds. It is an abstract entity created to describe an absence of value as perceived from a human perspective relative to the topic being discussed. Some think of zero as a threshold below which everything is different.

It's easy to count distinct things like people, apples, and money. What about chemicals and the risk of adverse health effects after exposure to them? Does the absence of a chemical in food, termed a *residue,* mean there is no detectable chemical using the best methods that modern chemistry has to offer, or does it mean just not enough chemical to cause you ill effects? Those ill effects, does this refer to sudden death, chronic heart disease, cancer after lifetime exposure or just a slightly upset stomach after a single meal? *These problems arise with measurement when one looks into the specifics and tries to assess if the value of something is truly zero in the context one is interested in. This specific issue was the incentive for me to write this book.*

Even in fields such as finance, where the concept of measuring zero in the context of financial worth is consequential, it becomes more complicated once one moves from assessing wealth by counting coins or paper currency to estimating market values of real estate, antiques, or complex securities and other financial instruments that depend on the state of the market and economy when they are eventually cashed in and counted. Zero in this case is the placeholder, much as in the number line, that divides one's net worth as having assets *versus* being in debt or even on the verge of bankruptcy. Such monetary transactions and asset counting in trade were the original driving force behind the development of numbers by ancient civilizations.

I have spent four decades in universities teaching and doing research in pharmacology, toxicology, nanomedicine, and biomathematics. I have repeatedly encountered the concept of zero and dealt with it in my research, taught and tried to explain it to students, and interpreted and debated its meaning to regulators. I served on numerous science advisory and expert panels for organizations including the Food and Drug Administration (FDA), Environmental Protection Agency (EPA), National Toxicology Program (NTP), United States Pharmacopeia (USP), and the National Academies of Science, Engineering and Medicine (NASEM). It is astonishing how broad the disciplines and issues are that require a proper definition of zero for the problem being considered, yet, how insular or siloed each individual agency or organization could be due to their own internal traditions or even biases. This has serious repercussions when a regulatory panel introduces zero in a regulation, or a country establishes a zero residue for trade, which when implemented may have severe unintended consequences. One example I encountered was the presence of a chemical contaminant in the water supply of a large city in Florida that was deemed too low to be of any consequence to humans consuming it. However, dairy cattle exposed to this same water, now have milk with the same concentration of this contaminant, but because of government regulations that demanded zero chemical residues in milk, the dairy could not sell their products, despite the fact that consumers would drink much greater volumes of water than milk!

My involvement in a United States Department of Agriculture (USDA) funded university consortium named the Food Animal Residue Avoidance and Depletion (FARAD) program, which I co-founded in 1981, has kept the concept of zero on the forefront of my professional life. FARAD's task is to calculate the time it takes for drugs and chemicals given to food producing animals to clear their bodies to a *no-effect* level for the human consumer, *e.g.*, when do they reach a very low or even zero concentration and are considered "residue free" in their edible tissues? Veterinarians use this resource to ensure that products derived from their food producing animals, be it meat, milk, eggs or even honey, are safe to eat. This concept of low residues can be well-defined by a toxicologist. However, very low concentrations of chemical residues may still be detected below the levels of concern by using improved tissue sampling techniques coupled to the extremely sensitive tools that modern analytical chemistry has to offer. When I first started working in this field over 40 years ago, the *allowable safe concentration*, termed a *tolerance*, was very close to the analytical limit of detection. Now, advances in chemistry detect compounds orders of magnitude below any possible biological effects, and below regulatory tolerances established years earlier when the drugs or pesticides used were first approved. Confusion arises because in some cases, minimal levels of certain compounds may have adverse effects on certain individuals (*e.g.*, think of allergic reactions to penicillin). Yet we are constantly exposed to very low levels of numerous natural compounds in our daily lives that may have no significant effect because our biological defense systems, honed by evolution, allow us to easily eliminate or metabolize these compounds to a harmless product.

Problems often arise when consumer groups conduct market surveys of food products and look for and report on *detectable, not violative*, chemical residues. This is where analytical chemistry, toxicology, regulatory, and trade policies clash. Concern is often greatly amplified through social media coupled to the recent public aversion to trust the opinions of experts in the field. Today, credentials for being an expert are often not academic qualifications, professional certifications or even scientific experience; rather the number of followers that influencers have on social media platforms seems more important. Such headlines of "chemical contamination" can go viral with traditional news outlets often joining the fray. As will be repeatedly argued in this book, *analytical zero does not equate to a zero biological effect and vice versa. Analytical detection and biological effects describe two distinctive domains.*

In addition to such misconceptions, what also prompted me to undertake this project, was my involvement in artificial intelligence (AI) and machine learning (ML) projects, where data is "scraped" from the open web and then used for calculations out of the context from the study that generated the data. I have seen large data studies where "0" simply appears in a dataset because there was no value entered. Calculations from these data then crash. As will be discussed later, many calculations and modeling studies performed in science use a logarithmic transformation of the data being modeled. If you try to take the logarithm of zero, the computer freezes. The author has even seen error message such as "The logarithm of zero is undefined because it is not a real number". Experienced workers are aware of this and curate the data into a database before analysis. Naive analysts are not so fortunate. With the very recent release of generative AI programs (*e.g.*, ChatGBT, Bard, Llama, *etc.*), scraping zero values from the web and using them as meaning zero could provide misleading and even dangerous conclusions unless the data is first curated. Specifics of these issues will be discussed in the chapter on mathematics.

Table 1.1 lists all the definitions and synonyms for zero I have encountered over my career and especially in writing this book. These definitions cover disciplines ranging from mathematics to chemistry to philosophy and will be expanded on in the subsequent chapters. Their definition is related to the context that they are employed, the trouble with zero occurring when it is applied across different contexts in different disciplines. Some of my scientific colleagues are so focused on their specific fields that they do not realize zero is operationally defined in the context of a specific discipline and may be different than what they are used to working with. Many regulatory committees are required to use the definition consistent with their regulations even when the data may be obtained from a different agency with

Table 1.1 Definitions and synonyms of zero

The arithmetical symbol 0 denoting the absence of all magnitude
No quantity or number
Nothing at all, not anything
The absence of any or all units under consideration
The absence of a measurable magnitude or quantity, non-detectable
The number dividing the set of all negative and positive numbers
The only integer that is neither positive nor negative
The placeholder in our number system used to reflect changes in order of magnitude: 0.01, 10, 100 etc.
Member of binary digit (0,1)
The identity element for addition
Zero-order rate process
Something arbitrarily or conveniently designated as zero, such as a size 0 dress
The empty set
A state of total absence
Chemical free
The lowest point or nadir
The lowest possible amount or level
The point at which a graduated scale begins (*e.g.*, thermometer)
The starting point of an event
A singularity, Janus point
An insignificant person or thing
The first person to be infected in an epidemic
To erase all official records of one's identity
A nonentity
A firearm sights set on a target
Adjusting an instrument to zero
Ground Zero—the center or point of origin of destruction
Blast off
Patient Zero—the first individual who sparks an epidemic
Limited visibility in a cloud ceiling
Numerology: potential and possibility
Synonyms: blank, cipher, empty, nada, naught, nil, none, nothing, null, void, zilch, zip

a different definition! I have repeatedly encountered this frustrating scenario. The approach to define an *operational definition* of zero is crucial to ensure such misuse doesn't occur. Yet this operational definition rarely gets permanently associated with the data for which it was defined. There are no meta tags attached to data elements providing a definition that would stay attached because zero has now moved and entered into a calculation for which it was never intended. When interdisciplinary issues arise, confusion often follows. A few examples will be introduced here but will be fully considered in later chapters.

The concept of zero has very specific connotations in many different scientific disciplines. An obvious case is the meaning of 0° temperature. We have been taught that its meaning is a function of what temperature scale we use. In physics, absolute zero is the lowest temperature theoretically possible where particles are essentially motionless:

- 0° Kelvin (K)
- −273.15° Centigrade (C)
- −459.67° Fahrenheit (F)

It is the bottom of the thermodynamic Kelvin temperature scale and well below the 0° of the Centigrade and Fahrenheit scales. As this comparison shows, the definition of zero degrees is very dependent on the scale used to measure it. In the Centigrade scale, 0 °C is the temperature where water freezes, whereas in the Fahrenheit scale it is the temperature at which a brine solution made of water, ice, and ammonium chloride freezes. Water freezes at 32 °F. The various temperature scales also nicely illustrate the utility of zero as a placeholder dividing negative from positive temperatures, except for the Kelvin scale where there is no temperature below 0°K (Fig. 1.3).

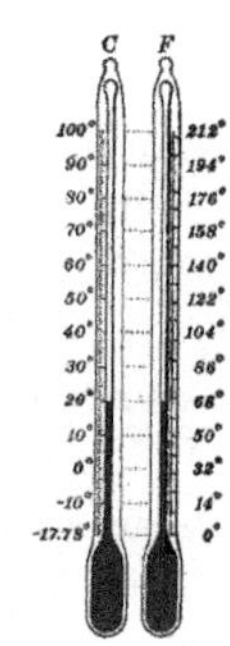

Fig. 1.3 Centigrade and Fahrenheit thermometers show equivalent temperatures

Physics defines the smallest distance possible in space as the Planck length (ℓP) equal to 1.6×10^{-35} meters. This leads to a definition of *zero distance* as being smaller than what some would regard as the granularity of space. Similarly, the shortest time interval possible between two events is the interval of time needed for light to travel one Planck length (t_P), which turns out to be 5.39×10^{-44} s. Anything shorter could be considered instantaneous or *zero time*. In cosmology, zero could also mean the so-called *Royal Zero,* which was the beginning of time when the universe was a singularity before the big bang. Zero in cosmology also denotes the volume of a black hole or a singularity, as it has no volume with a density approaching infinity. Ground zero is the site of a nuclear explosion. So many definitions even in the same field!

Similar confusion arises when one studies the field of chemistry. Zero is usually interpreted as the absence of a chemical that is not able to be detected by an analytical method or procedure. Of course, as seen in zero sugar soda, zero also simply means no sugar was added independent of any chemical analysis. The actual concentration present would be a function of the efficiency of the method used to extract the chemical from a matrix coupled with the physical or chemical property used to detect a chemical (*e.g.*, light absorbance, mass, charge, *etc.*) and the sensitivity of the detector used to measure its concentration. In the field of analytical chemistry, assays in the 1950s used colorimetric, spectrometric, or microbial bioassays to detect drugs, such as antibiotics in food or milk. Their lower limit of sensitivity was approximately one part per million (ppm—1 mg/L). Improvements in extraction techniques and chromatography systems pushed these down to one part per billion (ppb—1 μg/L). Mass spectrometers and other tools of modern analytical chemistry now can detect chemicals at the level of parts per trillion (ppt—1 ng/L or 1 pg/ml) and lower. A ppt is equivalent to one drop of water in an Olympic size swimming pool. Confusion arises when interpreting the difference between a limit of detection (LOD) *versus* the limit of quantification (LOQ). As will be fully discussed in this book, a zero concentration in 1960 is *very different* from a zero concentration today. Yet some regulations that are active today still use the definitions formulated six decades ago.

This concept of detection limits is nicely illustrated by the simple example of using a microscope to examine a drop of water. When looking at a sample of water with the unaided eye or even at low magnification of an optical microscope, nothing may be seen. At that magnification, one would be correct in stating "nothing was present". However, as magnification increases a particle appears, which at higher magnification could yield a bacterium. Simply shifting magnification, and thus physical scale, changes our ability to detect substances. This concept of the *limit of detection* will be crucial in the practical application of defining the absence of a substance, because instrumental extensions of all of our human senses open up new scales of nature previously hidden from view (Fig. 1.4). A similar concern today exists with the detection of microplastics in water. A similar concern today exists with the detection of microplastics in water.

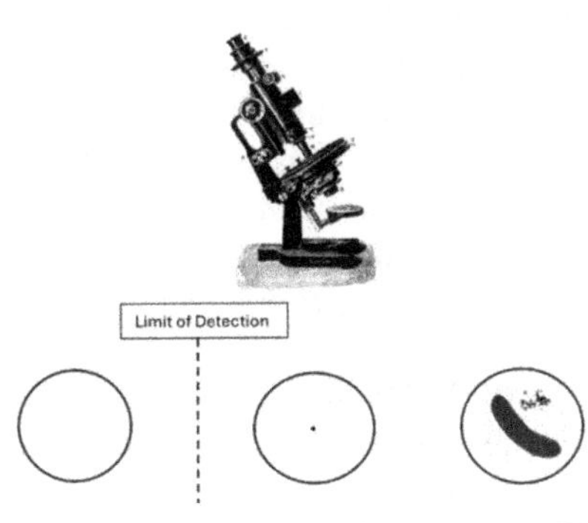

Fig. 1.4 Illustration of how one's perception of nothing changes when instruments magnify the limit of human sight

This figure is similar to a controversy a few decades ago where water in New York City was judged to be *unkosher* by the Orthodox Jewish community because microscopic examination discovered copepods in their drinking water, requiring filters to be added in homes to remove them (*J Berger, New York Times, November 7, 2004*). At that time, New York municipal water was exempt from federal standards that require filtration. The religious dilemma was that copepods are a crustacean distantly related to shrimp which are prohibited by the Orthodox Jewish faith. Although most were invisible to the unaided eye, they are present world-wide in drinkable water and present no health hazard, their presence now known was deemed prohibited. A Talmudic rabbi explained that *hidden things belong to God and man is only responsible for what we see. If one doesn't see it or know about it, then it doesn't exist*. This definition nicely reflects the root of the problem of defining what *nothing* really means!

The same progression of technological advances in measurement also applies to the incredible accuracy of modern scales for weighing substances down to fractions of a milligram. If we still measured in stones as the ancients did, zero weight would be anything less than about 6 kilograms or 14 pounds. Similarly, in audiology, zero decibels are defined at the point which humans can hear sound; a value that does not extrapolate to other species nor to humans with modern assisted hearing devices. As one can see, the meaning of zero in space, time, weight, volume, or concentration changes as the scale changes and technology advances.

A primary focus of this book will be studying how to interpret zero in the context of biological effect. There are multiple levels to this issue. Does the presence of a chemical cause damage to or kill an individual cell, affect function in a vital organ, such as the liver or kidney, harm the whole body, or cause potential damage to a population or even an ecosystem at large? These zeros are very different. So-called *in vitro* assays expose individual cells grown in culture to chemicals to assess potential biological effects. However, in live intact animals or humans termed the *in vivo* setting, a chemical may never be able to reach such a target cell after exposure in the food or the environment. Individual cells are often protected by other structures and elimination organs (*e.g.*, liver and kidney) which were designed by evolution to shelter cells from exposure to outside chemicals and eliminate thise hazardous substances that have entered the body.

What are the criteria and sensitivity of a method to detect an adverse health effect? Is one interested in acute illness after a single exposure or chronic disease after lifetime exposure? What happens when a safe level is defined using what is now outdated chemistry, or an insensitive or inappropriate toxicological test? In toxicology, this has led to the development of numerous concepts, such as *No Observed Adverse Effect Level* (*NOAEL*), *Lowest Observed Adverse Effect Level* (LOAEL), *Maximum Tolerable Level* (*MTL*), *Threshold of Toxicological Concern* (*TTC*), *De Minimus Level,* and *Threshold of Regulation*, all of which will be defined in later chapters. What does *Net Zero Emissions* in environmental air quality regulations really mean and how is it calculated?

An area where interpreting zero has attracted considerable public health and medical researchers' attention is the extrapolation of the dose of a potential carcinogen to zero. This activity has been a driver of chemical and drug regulations worldwide and has resulted in specific legislation, such as FDA's Delaney Clause (1958), which prohibits the use of any additive in food that has been linked to cause cancer in animals or humans. Issues of zero abound ranging from what concentration of a carcinogen causes cancer in humans to models of how to extrapolate and model a chemical's propensity to cause cancer from laboratory animal or computational studies. Well-regarded scientific research suggests that according to the theory of *hormesis*, low exposure of a toxic chemical to humans may actually produce a beneficial effect. Similar thinking underscores belief in the value of *homeopathy*, however the principles behind and validity of these two concepts are very different. There are debates in mainstream medicine as to "what is the criteria for zero brain activity reflecting death?" When zero appears in such clearly important debates, it is crucial that cross disciplinary definitions be respected and not misused. This is even more important to convey these definitions when results are then communicated to the public or incorporated into law. While we are talking about medicine, recall *patient zero* is the individual that epidemiologists believe is the origin of an infective disease epidemic.

These different standards for what zero means are particularly important when zero tolerance is established in a regulatory or trade policy. This can be establishing a tolerance of a drug in a food product by a regulatory agency for sale or import, or a private quality control standard required before purchase from a supplier. If zero residues were promulgated as a policy, say in 1980, I can assure you that a modern analytical chemistry technique will find something there today. Does this change in detectability affect the potential health effects to a person now consuming this food? If the concentration used in the original product approval studies indicated safety at that level, why worry now if the chemical can be detected in the body below this historic established safe level? Or has our ability to detect adverse biological effects now improved so that the tolerance needs adjusting because of biology and not chemistry?

It is well known that certain standards are used as barriers to international trade but do not protect human health. Similarly, drug tests are widely used in sporting events to detect the presence of illegal drugs or hormones, or in horse racing to detect doping. In these scenarios, the goal is detecting use and not biological effects. As will be discussed, these are issues involving the interplay of concepts and practices in chemistry and biology. Because they are all measurements, statistics also come into play. An interesting note is that the FDA's current strategy (released April, 2021) to reduce chemical contaminants in food is named "Closer to Zero", reflecting both the inability to actually reach zero and that zero remains in the vernacular as the standard of purity in foods. It would appear that when zero is used in the context of no risk or absolute purity (*e.g.*, food with no contaminants, chemical-free), zero takes on a positive connotation of perfection.

A final problem with zero relates to its special properties in mathematics. As mentioned, zero occupies a unique place in our number system as an integer not representing any value. Due to some interesting properties that will be fully discussed in a later chapter, if zero finds itself in the denominator of a function or as the exponent of a number (Y), strange things happen. We know we cannot take the logarithm of zero. Division by zero ($Y \div 0$) is not possible. Multiplying by zero erases the value of the number being multiplied ($Y \times 0 = 0$). Zero factorial equals one ($0! = 1$). Raising anything to the power of zero results in the expression having a value of one ($Y^0 = 1$), except if it is zero being raised to the zero power (0^0), which is also not possible.

Calculus developed the concept of *limits, where the value of a function is estimated as it approaches zero*. When complex mathematical models are used or software programs link to databases to input data, a value of zero may appear in an equation for all the reasons discussed above and cause the calculation to crash. This can happen when a concentration is automatically entered into an equation as a divisor, a denominator, or an exponent. This is where the concepts of LOD and LOQ come into play to deal with the occurrence of very low concentrations. By using LOQ as a divisor, say to estimate the time for elimination of a drug from exposure to "zero" concentration, a value can be computed but will vary greatly depending on the final concentration selected.

Zero also has a strong presence in statistics. When low concentrations are extrapolated to estimate values, issues occur when a mean concentration may be above zero but the lower bound of a confidence interval includes it. The concept of zero is fundamentally embodied in the statement of the null hypothesis, a central tenet of statistical hypothesis testing. *Null is defined as having or associated with the value of zero.* The null hypothesis is the baseline value of a quantity, function, or proposition that is accepted as the standard against which one's new hypothesis is tested. This new hypothesis could be thought of as an educated guess about something in the world that you can quantitate. In science, it's the experimental relationship that you want to prove is superior to the existing relationship embodied in the null hypothesis. This concept of zero expressed in the embodiment of the *null hypothesis* holds a central role in statistical theory.

Statistical testing involves estimating the probability that the null hypothesis, generally the existing estimate or scientific hypothesis, is wrong. A statistical test assigns a level of confidence that the *null* hypothesis (H_0) remains true against the new data being assessed. An example would be testing whether the length of time it takes to eliminate a drug under new conditions like age or disease (H_1), is greater than normally expected (say for this example 10 days). The two hypotheses would be stated as:

H_1: Time > 10 days
H_0: Time $\leq$ 10 days

Data is then collected across several individuals noting whether they are healthy or for example have a certain disease, or are in a specific age range (e.g., < 5 years old or > 65 years old). Depending on the nature of the data being analyzed, a statistical distribution is assumed for the data set (*e.g.*, normal distribution) and an estimate of confidence is assigned that reflects how much risk you are willing to accept that the null hypothesis may actually be true compared to the new hypothesis being tested. This is termed a Type I error [rejecting H_0 when H_0 is actually true] with a probability or p-value of α, normally assigned to be $p \leq 0.05$; meaning there is only a 5% chance (1 out of 20) that in fact the null hypothesis was true (in this case time was actually ≤10 days) and your data set or experimental result did change the status quo. The p-value is the estimated probability of this error occurring while α is called the *level of significance* of the test. In practice, the level of 5% is the maximum that will be used but can be much lower depending on the nature of the problem. We will go into more detail about statistics in a later chapter as there are many factors that come into play when doing statistical tests, including sample size, data distribution, and the importance of being correct (that is magnitude of α). However, at this point in discussing the basic meaning of null and zero, the concept of null and even the meaning of zero as the absence of quantity, are both centrally entangled in this field.

The word Null is also encountered in geography as the coordinates of 0^0 latitude (the equator) and 0° longitude (the prime meridian), the intersection points of these two orthodromes being referred to as the Null coordinates (0° N, 0° E). These great circles divide the world into Northern and Southern as well as Eastern and Western hemispheres. The location of the equator denoted as 0° latitude is defined by planetary geometry as the great circle equidistant between the North and South poles. The prime meridian denoted as 0° longitude was defined in 1884 as the longitudinal great circle passing through Greenwich, England, which then defines Universal Time (UT or UTC) as Greenwich Mean Time (GMT) where a clock located there strikes zero hours at midnight (00:00 UT). This geographical intersection point at the Null coordinates is now commonly referred to as Null Island, though no land mass is present at this location off the coast of Ghana in the Gulf of Guinea in the Atlantic Ocean. It is marked by a weather buoy (PIRATA Station 13010—Soul) permanently moored there in some 16,000 ft. of ocean water. With the advent of GPS systems, when an address is added to a map, a geocoder converts this physical address into geographic coordinates. If an error occurs, or if a GPS system is turned off, the coordinates are assigned as "0,0". The problem arises when naive software programs reading geographical data map this onto its location off the coast of Africa. An unbelievable number of addresses are located at Null Island! (Fig. 1.5)

Fig. 1.5 The physical *Prime Meridian* in Greenwich UK and the location of the theoretical *Null Island* at 0° longitude and 0° latitude

The null coordinates in the theory of general relativity in physics refers to where the spacetime interval between two events on the path of light rays is zero. As the reader can begin to appreciate, zero and null always appear to have central places in a discipline.

A lot of the angst and unique properties associated with zero originated from the philosophical concepts of emptiness, nothingness, and the void developed over centuries of western and eastern philosophy. The concept of nothingness arose simultaneously in the fifth century BC in ancient China, India, and Greece; embodied in the philosophies ranging from Taoism to nihilism. Zero was inextricably linked to the void and nothingness, beliefs that often engendered fear and even chaos. The Greeks had no zero in their mathematics nor did they believe the void was possible. Its concept bleeds into the interpretation of the fourth century BC Greek philosopher Zeno's famous paradox. It only started to abate when post-Socratic philosophers including the atomists Leucippus and Democritus, had to accept emptiness as the space between atoms, a concept carried forward in history by Lucretius who strongly influenced later natural philosophers. Even Aristotle did not believe that "0" nor "1" were true numbers, with his first number being "2". The concept of nothingness and nihilism continue in the modern and contemporary eras embedded in philosophies ranging from Buddhism, Hinduism, skepticism to existentialism, and post-structuralism. It seems the concept of nothing continues to have a strong hold on the human mind.

This can be appreciated from the origins of the simple word *naughty*. In the 1300s, the word naughty referred to people who had *naught* (nothing); they were simply poor or needy. However, by the next century, the meaning of naughty evolved from *having* nothing to *being worth* nothing, or alternatively being morally bad or wicked. It could refer to a promiscuous or sexually provocative person, or someone guilty of other immoral behavior. Thus, by the middle of the second millennium, "naught" the synonym for "nothing" took on a morally offensive connotation.

It may be that zero's stickiness to and persistence with such a negative connotation is because it became a cultural *meme;* an idea, concept, or behavior that according to the evolutionary biologist Richard Dawkins is passed along from generation to generation. They function as cultural analogues of genes, with individuals receiving them not consciously aware of the bias they could introduce. As will be seen in the next chapter, negative philosophical and religious connotations could further become attached to this zero or nothingness meme which subconsciously prevented its adoption in fields such as mathematics.

Zero still has a negative connotation for gamblers playing the roulette wheel since "0" is the only number where everyone loses, unless you bet on it. That is, "0" is not included in bets such as even—odd, or red—black. In European roulette, there is one "0" on the wheel, in American roulette (see picture below), there are two: "0" and "00", and in Sand's roulette, there are actually three, which is great for the house but not for the gambler! (Fig. 1.6)

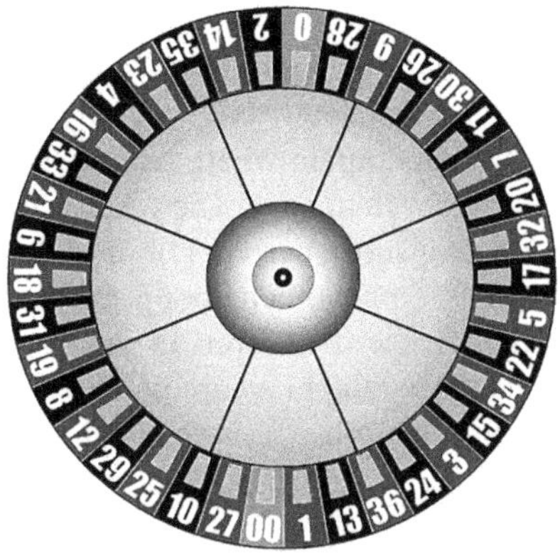

Fig. 1.6 American roulette wheel with 0 and 00 stops

The reader can now appreciate that defining zero and having discussions about nothing are far from empty pursuits. It is encountered in most areas of life. A perplexing one that has always confused me is what does a zero dress size really mean? As Ellen DeGeneres has said it could only stand for "*Ohhh my God, you're thin.*" Zero has a grasp on the human psyche that directly affects how its meaning is interpreted when applied to multiple fields. Finally, even the popular American sitcom *Seinfeld* was billed as "the show about nothing".

A simple solution when working with zero is to have an *operational definition* that is appropriate for the context of the topic at hand. This would then allow the reader or subsequent user of the data to know how zero was defined, and if that operational definition is appropriate for the task at hand. It is my hope that the more in-depth perspectives on zero presented throughout the rest of this book will help sharpen one's perspective on the meaning of and determination of zero.

Approach of This Book

As can be appreciated from the discussion thus far, there are many perspectives that yield different definitions of zero, empty, and nothing. The first six chapters will attempt to independently define zero within specific disciplines where there is some consensus as to its meaning. The journey begins with the ancient philosophers and theologians where the concept of nothing was first encountered and debated. This is followed by how the number zero worked its way into modern mathematics. The philosophy of nothing and the development of zero were closely entwined throughout most of history. These discussions will also show stark contrasts between western and eastern schools of thought. The chapter on mathematics and statistics will review properties and rules that govern the use of zero in calculations and the concept of the null in modern mathematics and statistics. This will be followed by the definition of zero in physics relative to mass, temperature, time, and sound. The chemistry chapter will move up the space scale to the world of atoms and molecules, the domain where most of the public understanding of zero arises, especially analytical chemistry. The next chapter of biology will change focus to go from not detecting substances to not producing biological effects through discussion on how these two phenomena connect. The remainder of the book will explore how these hard-science definitions of zero work their way into and often get distorted when used in more applied applications of toxicology, carcinogenicity, food safety, regulations, and other sociomedical domains.

Selected Readings

Dantzig T, Mazur J. *Number: The Language of Science*. New York, Plenum Press, 2007.
Dawkins R. *The Selfish Gene*. Oxford, Oxford University Press, 1976.
Kaplan R. *The Nothing That Is: A Natural History of Zero*. Oxford, Oxford University Press, 1999.
Matson J. *The Origin of Zero*. Scientific American, August 21, 2009.
Riviere JE. *Chemical Food Safety*. Ames, Iowa State University Press, 2002.
Seife C. *Zero—The Biography of a Dangerous Idea*. New York, Penguin Books, 2000.
Sorensen R. *Nothing—A Philosophical History. Oxford,* Oxford University Press, 2022.

Chapter 2
Philosophy

> The biggest questions in science and religion are about nothingness and eternity, the void and the infinite, zero and infinity. The clashes over zero were the battles that shook the foundations of philosophy, of science, of mathematics and of religion. (Charles Seife)

A discussion of the history and origins of zero benefits from an examination of the historical roots from where the concept of emptiness and the void arose. In my readings and research leading to this book, I was intrigued by the speculations presented as to why there was such a lag in the development of the digit zero amongst ancient civilizations, considering the relative sophistication of some of their numbering systems as well as the accuracy of their astrometric forecasts. I am the first to admit that I am not a professional philosopher nor historian, and thus I will tread lightly. For readers wanting a more in-depth discussion of these philosophical topics, I suggest reading Roy Sorensen's *Nothing—A Philosophical History*.

The concept of *empty space*, *void*, and *nothingness* occupied the minds of many early scholars and philosophers. In many ancient cultures, *there was a primal fear of the void and chaos, and thus a fear of zero*. In the West, there were strong theological pressures to avoid dealing with the existence of empty space. Tracing the historical roots of such concepts is extremely difficult because the historical record is incomplete, often compiled in different archaic languages, prone to bias introduced by ancient scholars and translators, and most importantly, the likely direct destruction of *heretical* material that went against the prevalent dogma of the time. When we fast forward to the modern era, the concepts of nothingness and emptiness exist as core components of many philosophies, including most Eastern religions, existentialism, and nihilism. Therefore, a discussion of this concept, which may have affected the emergence and interpretation of the number zero, still triggers contemplation in many modern thinkers, and warrants a brief review.

Figure Use Attributions
Figs. 2.1, 2.2, 2.3, and 2.4: Wikimedia – No restrictions
All other figures are either self-drawn, photographed or from author's own publications (attributed)

J. E. Riviere, *Zero – Much to Do About Nothing?*,
https://doi.org/10.1007/978-3-031-82998-7_2

The Ancient Greeks and Romans

Two names stand out when one contemplates the meaning of emptiness in ancient Greece: *Leucippus* and his student *Democritus*. Both were believed to have lived as contemporaries in the fifth century BCE around Abdera (debate exists on exactly where both lived), though they were younger than their other contemporary, *Socrates*. They are widely recognized as natural philosophers, which at that time also involved the study of astronomy.

Leucippus was a student of *Zeno of Elea*, who in turn was a student of *Parmenides*, the founder of the *Eleatic* school of philosophy. Parmenides was probably the earliest philosopher to contemplate and make pronouncements on emptiness, suggesting that the void did not exist. Zeno was noted for various *paradoxes* including his most famous being the *Bisection and Dichotomy Paradox*, studied by countless philosophers over the centuries. He postulated *motion was an illusion due to each next movement being infinitely divisible, thereby preventing any forward progress in time to occur.* This is manifested in the more popular Achilles and the Tortoise paradox. In reality, the solution to these zero-time interval dilemmas ultimately was solved with the invention of calculus millennia later. Some have speculated that atomism and the concept of the void also grew out of contemplation of these arguments.

Leucippus gets most of the historic credit for the theory that the universe consists of two fundamental elements: the *full* or *solid* and the *empty* or the *void*. Both were thought to be infinite. Democritus further expounded on atomism and for our purposes, empty space and the void. Atomism has been postulated to have arisen from the Eleatic belief that *what is* must be one and unchanging, which makes the corollary of *what is not* an incomprehensible proposition from this perspective. This engrained belief in early Greek philosophy may have even prevented the concept of *something coming from nothing* from ever being considered. Leucippus, in the formulation of atomism, also held that atoms were always in motion which required a void for them to occupy.

Democritus who travelled extensively through the ancient world (*e.g.*, Egypt, Persia, India, Ethiopia), was a polymath, authoring works in numerous fields, including ethics, music, physics, cosmology, and mathematics. Democritus expanded on the concepts of Leucippus viewing the universe as composed of the indivisible atoms that move about in an infinite void, with the unoccupied regions between atoms comprising the void, rather than there being an absolute encompassing space exterior to where the atoms existed. A central part of this philosophy was that change did not require that something arose from nothing, change instead is a rearranging of the fundamental atomist building blocks. Macroscopic objects we see are but rearrangements of these indivisible particles. They strongly believed, in response to Zeno's Paradox, that there was a lower limit to divisibility. While having to accept the void, they also had to accept the apparent contradiction of *what does not actually exist*. Democritus was also reported by Plutarch to have worked in mathematics and was aware of questions concerning the relationship of his view of atoms and the void relative to the nature of mathematical geometric solids.

Relative to our discussion of the impact of emptiness and the void on the human psyche, *Plato* argued against the existence of only matter and the void. He held, like Parmenides, that there was no void. All motion is produced in an absolute *plenum* (no spaces empty of matter) and therefore took what is referred to as a *vortical form*. He also believed in an incorporeal world of absolutes where we live, where we only view the *shadow* world that is defined by the Platonic absolutes and the Forms. There is obviously so much more to Plato's philosophy, but for our purposes, the belief that there is no void and that a non-material ephemeral world exists could help in speculating that in this world view, there was no need to quantitate something that simply could not exist.

Epicurus, a contemporary and antagonist to *Aristotle* who lived from 341 to 270 BCE, was strongly associated with atomism as well as atheism. Today, he is remembered primarily for his hedonistic philosophy of Epicureanism *that the goal of human life is happiness, the absence of physical pain and mental distress.* He also prescribes how an Epicurean should approach a problem; *first gather evidence and talk to trusted persons with first-hand knowledge of an event, trust one's senses and test assumptions.* His natural philosophy was a direct outgrowth of atomism. His writings accept the void as being both infinite and without substance. He considered the smallest and indivisible parts of atoms as the limits of length, a concept consistent with the modern Planck Length of quantum mechanics, which will be discussed in more detail in a later chapter. An Epicurean felt *death is nothing to us, for what is dissolved back to their original atoms has no feeling, and that which has no feeling is nothing to us*. This stance on the lack of life after death coupled with his atheism was attacked by his critics. By discrediting these beliefs, a failure to adopt other aspects of his philosophy and methods occurred, thereby losing for centuries the concepts of atoms, the void, nothingness and what appeared to be the rudimentary beginnings of the scientific method (Fig. 2.1).

Fig. 2.1 Epicurus

The *Stoic* school, formed around 300 BCE, was a competitor of Epicurus. A prime tenet was corporealism, the belief that only bodies exist and have the capacity to act or be acted upon. The four elements comprising the cosmos, first proposed by the philosopher Empedocles who considered them the fundamental building blocks of the universe, were fire, air, water, and earth. Like the atomists, void is essential as it allows bodies to move but lacks the capacity to act or be acted upon. Stoics classify void as incorporeal, something rather than nothing, even though it does not exist.

Aristotle, living from 384 to 322 BCE, studied in Plato's academy, tutored and mentored Alexander the Great, and became one of the most well-known and influential philosophers in history; having shaped much of Byzantine and Western Christian thought up through the Renaissance. Because of this widespread impact covering almost two millennia, many secondary accounts of Aristotle's teachings are biased by various secular and religious translators and traditions. He held, as did Plato, that God was incorporeal, and his providence extended to the heavenly bodies. Aristotle postulated three primary classes of substances comprising the cosmos: form, matter, and the compound of form and matter. These substances are the basis of the universe and everything within it. Rules only applied to like substances. For example, mathematics would not apply to real substances. There were five elements making up the cosmos: fire, air, water, earth, and ether; the latter occupies what was in space above the Earth.

To Aristotle, the void (τὸ κενόν) was impossible, a belief that is clearly reflected in his famous quote: *Nature abhors a vacuum* (Book IV of Physics). Empty space is a logical impossibility. This belief that empty or unfilled spaces are unnatural, because they go against the laws of *nature,* significantly influenced Western thought for a millennium. It may have been here that the negative cultural meme associated with zero first took root. Matter was infinitely divisible and filled all of space in what was the plenum. The concept of ether or plenum filling empty space remains prominent in many later philosophies and scientific theories well into the modern era. It clearly contrasts with the atomists. Consistent with this aversion to zero, Aristotle's numbering (*arithmos*) system had no "0", but also had no "1" as he postulated that "1" could not be both the precursor of the numbers and not possess a precursor to itself. He held that since a number is a plurality of objects, "1" is not a number. His counting numbers thus began with "2".

There is a common theme that emerged from the ancient Greeks relative to the concepts of the void, non-existence, and emptiness. Philosophers such as Leucippus, Democritus, and Epicurus, were developing what one now could consider relatively modern views of the relationship between something and nothing, atomism, an exploratory deliberative process based on observation and testing of assumptions, as well as the acceptance of a void. In contrast, the prominent Greek philosophers Aristotle and Plato found any void within the cosmos as unacceptable. It is the latter's views that became widely adopted and served as the foundation of Western and especially Christian thought. Aristotle surfaced as the primary authority of Christian theology. This underlying philosophy that rejected the concept of the void and thus atomism, coupled to Christian belief in discovering the truth only through holy revelation, not observation or experiments, became the dominant framework of natural philosophy for the next almost two millennia.

The philosophies prevalent in ancient Rome were a continuation of the earlier Greek thinkers and in many cases, much of what we know about the Greeks were from these later Roman writers. The only known work of the Roman poet and philosopher *Lucretius,* who lived from 99 to 55 BCE, was *De Rerum Natura* (*On the Nature of Things*), a work on Epicurean philosophy. The work influenced other Roman poets including Horace and Virgil and clearly expounded, in rich poetic language, the Epicurean tenets of atomism and cosmology. The world consisted of atoms and the void, each being complementary. He believed that the universe arose by chance and not by divine intervention of the many Roman Gods. Unfortunately, his work was lost or suppressed throughout most of the Middle Ages and only reappeared in Germany, some say, in the early fifteenth century [there is debate on when his work actually resurfaced, but to our point, it was suppressed and not really read till the Renaissance].

Cicero was a Roman orator, statesman, and philosopher that in various writings followed disparate philosophers ranging from Epicurius to Aristotle and Plato and even to the *Skeptics*. A contemporary of Julius Caesar, living from 106 to 43 BCE, he is often regarded as a preacher of Stoicism to at least the Roman elite and was involved in the transformation of the Roman Republic to the Roman Empire. Concerning our interest in the void, the phrase *nothing can be something* summarizes the Stoic theory of the void. The void has elements of it being both something as well as just empty space defined as the absence of matter. To them, the void can only exist if matter also exists. By the end of the Roman period, active debate and uncertainty as to the meaning and true significance of the void faded as Christian religious dogma, heavily influenced by the teaching of Aristotle and Plato that asserted that the void did not exist, dominated Western scholarly activity well into the Middle Ages.

Eastern Philosophy: China and India

Compared to Western philosophy, ancient Chinese philosophy has a rich history of contemplating emptiness and the void, making the acceptance of this concept a significant difference between Western and Eastern philosophies, religions, and even art.

Many scholars reviewing these works fully admit that oversimplification and erroneous linkages of ancient events can lead to conflicting conclusions. For example, Confucianism is as much a philosophy as it is a system of political ideas. Its origin is debatable but is considered largely based on the teachings of *Confucious* who lived from 541 to 479 BCE. His writings are obviously considered its foundational works, but the writings of the earliest ancient Chinese "Confucian" philosophers were not uniform; a similar state that we saw with our knowledge of ancient Greek philosophers. As occurred with ancient Western work, early texts are often reinterpreted by later scholars with different and often biased perspectives. Because these philosophical and religious concepts have deep and varied meanings, whose interpretations are not central to the focus of this book, they will only be briefly introduced so that the reader can get a flavor of how concepts of nothing and the void have permeated the human psyche since ancient times. These different concepts no doubt contributed to fundamental differences in the sciences and cultures that emerged from these civilizations in modern times.

In Chinese *Taoism*, X*u (虛)* connotes emptiness or purity, an equilibrium state where one is open to a transforming experience (*Tao*). The Taoist void is the empty incipience that harbors all potentialities. In *Taoism*, Tao is the non-empty void through which Tao flows in a person, with *Tao* being the Great Principle, the Creator and Sustainer of everything in the universe. The state of emptiness (*Wu*無) is viewed as a condition of stillness and placidity, the state of one's pure mind and a mirror to the universe, a positive state sought after through meditation. This is very different from that of the Greeks where when they look to the smallest scale of matter, they contemplate and even fear the emptiness of the void. It's difficult to trace the impacts of these different mental perspectives on society or for our purposes, the development of zero. The first Taoist philosopher was believed to be *Lao Tzu* living around 500 BCE, an entity that many now believe was the pen name for numerous sages studying and writing in this area. The religion of Taoism (also known as Daoism) more formally began in a revelation to *Chang Tao-ling* in 142 CE. A great example of these philosophical concepts was the pronouncement of Lao Tzu that *clay may be shaped into a pot, but it is the space inside that is useful.* Similarly, *windows and doors are cut into a wall, but it is these holes that are useful*. In these examples, the void is the functional unit. Philosophically, Lao Tzu praised the virtue of inaction and minimizing one's self, instead letting energy or the Tao takes its course. One's baseline for comparison in all activities, including ethical considerations, is always inaction.

In Buddhism, similar to Chinese thought, emptiness (*Sunyata* in Sanskrit, *Ston-pa-nyid* in Tibetan and *Kung* in Chinese*)* can be translated as void or openness, with concepts close to the base meditative states seen in Taoism. Emptiness in Eastern mysticism is the state of pure consciousness in which the mind has been rid of all objects and images. *Confucianism* makes us appreciate the outline of various matters, but not the details of their content, creating an ambiguous space that contributed to the origin of the concept of emptiness that emerges in Taoism and Buddhism. In Confucian tradition, emptiness is opposed to fullness, whether defined as a void in physical or metaphysical space.

These philosophies and religions are closely linked through this concept, a concept that is also fundamental to Chinese aesthetics. For our purposes, these schools of thought should be perceived as different dimensions of a single Eastern philosophy that continues to influence Asian civilization thousands of years later.

Empty space in Asian art represents the concept of the void or emptiness, so-called *painting the white*. Its centrality to Asian aesthetics is illustrated by comparing a European garden, which typically is a dense arrangement of flowers, vegetation, and statues, versus a Japanese one, where open and empty space may be its focus. Finally, as an illustration of these concepts of what is emptiness *versus* existence, consider an optical illusion such as *Rubin's Lady and the Vase*. Which exists, the white vase on a black background of emptiness, or silhouettes of two women staring at each other on a white background? (Fig. 2.2).

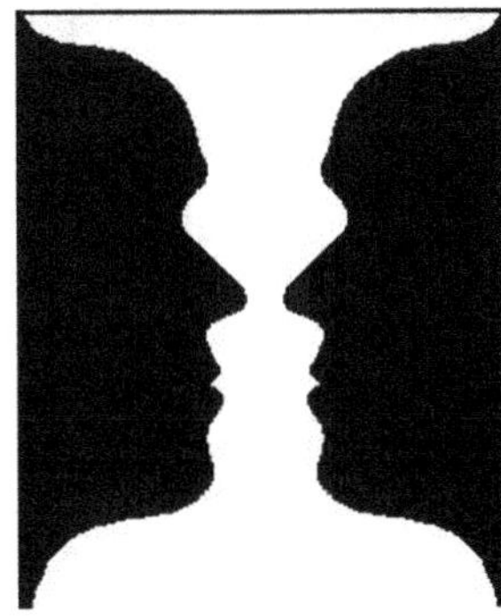

Fig. 2.2 Rubin's vase optical illusion nicely illustrates one philosophical concept of emptiness

A discussion of Indian philosophy in many cases overlaps that of the Chinese scholars due to the strong influence of Buddhism. The Buddist monk *Nagarjuna,* living from roughly 150 to 250 CE, believed if the existence of each object depended upon the existence of another, then there would be nothing grounding the entire chain of existence. However, Hindu philosophers are crucial in understanding the concept of nothingness and probably the first emergence of zero. In *Hinduism*, emptiness or nothingness relates to the vastness that is the source of creation.

In the earliest age of the gods, existence was born from non-existence—The Rig Veda.

Everything comes from nothing and eventually returns to nothing. Nothingness existed before God undertook creation, it is greater than creation which comes and goes as worlds appear and disappear in cycles. According to this theory of emptiness, any belief in an objective reality grounded in the assumption of intrinsic, independent existence is untenable. All objects and events, including abstract concepts, such as time, are devoid of an objective and independent existence. The Indian *Vedas* holds that *nothingness is the ultimate source of knowledge*, the final human frontier beyond which there is nothing to be known. Far beyond the scope of this book, one can appreciate differences between Hindu and Buddhist concepts of emptiness and nothingness compared to their Western counterparts. However, it is clear that different and varied schools of ancient Indian philosophy actively considered, debated, and meditated on the meaning of nothingness and the void.

It is not surprising then that zero as a number first emerged in India in the fifth through ninth century CE, a millennium before this "discovery" finally emerged in Europe. This will be fully discussed in the next chapter where the history of the number zero will be dissected. *The focus here is to explore how the philosophical background in different cultures may have facilitated or inhibited its ultimate emergence.* Moreover, closer to the primary focus of this book, the *Taoist* philosophical proposition *doing nothing may be the best action* in certain scenarios; leads to the foundational concept underlying *testing the null hypothesis* in statistics, and only rejecting it when there is overwhelming evidence of the validity or accuracy of the new hypothesis being presented. Similarly, the toxicological risk assessment strategy embodied in *The Precautionary Principle*, the cornerstone of modern European Union's risk tolerance regulations, could be considered a Taoist proposition, where if risk of an adverse effect exists, the best path forward is to avoid it.

Europe and the West

A history of nothingness and the void in Medieval Europe (approximately 500—1500 CE) is largely a rediscovery and reemergence of the ancient Greek and Roman philosophies. Many of the texts were translated from Greek, Latin, and Arabic and kept and preserved during the Dark Ages within Catholic monasteries and later cathedral schools throughout Europe. Many texts were probably deliberately destroyed as being heretical writings. Some scholars suggest that even as late as the turn of the first millennium, superstition prevented any so-called "*digit of the void and evil*" (concepts referred to in the Bible as none, nothing, empty, nought, and void) to be widely adopted in Europe until 1200 to 1300 CE, and for some cases even as late 1500 CE, despite it being used in India and Persia in the last half of the first millennium. This strong negative view of the void and emptiness can be appreciated by examining biblical passages where the void is linked with darkness, evil, and the lack of God's presence.

- *And the earth was without form, and void; and darkness [was] upon the face of the deep. And the Spirit of God moved upon the face of the water.* (Genesis 1:2)
- *She is empty, and void, and waste: and the heart melteth, and the knees smite together, and much pain [is] in all loins, and the faces of them all gather blackness.* (Nahum 2:10)
- *Let him not trust in emptiness, deceiving himself, for emptiness will be his payment.* (Job 15:31)

Similar associations can be found in Byzantine writings associating the void with blackness and evil. It is little wonder that texts discovered by religious orders that were contrary to this dominant belief never saw the true light of day.

The study of the history of medieval philosophy has undergone a change in perspective over the last few decades. Classically, scholars focused on Latin texts from Western Europe. Recently, the discipline now includes Islamic and Jewish scholars writing in Arabic and translated into Latin, Greek texts written by Byzantine Christians, and Jewish scholars writing in both Arabic and Hebrew. As the reader can appreciate, this infusion of cross-cultural sources nicely illustrates why the precise historical mapping of where the concept of zero originally emerged is buried in the translations and their interpretations from multiple sources and beliefs across Europe and the Middle East. It's also interesting to note that despite the diversity of scholars who contributed to the corpus of what is now referred to as "Western" philosophy, it remains largely based on original Greek and Roman texts in contrast to philosophies that arose in India and China. As we will see in the next chapter, once zero emerged during the Renaissance in Europe, the development of our modern numbering system, calculus, and modern science soon followed.

The hallmark of medieval Western philosophy, and thus Christian beliefs, was based on the work of Aristotle, and to a lesser extent Plato. Albertus Magnus and Thomas Aquinas were largely responsible for reintroducing Aristotle to the West and reconciling his philosophy with Christian thought. The study of Aristotle, or that of specific commentators' versions of Aristotle, such as those of Avicenna, Aquinas, Augustine or Peter the Lombard, was central to most higher education until the late seventeenth century. Because of this philosophical heritage, concepts such as the void, nothingness and emptiness fell on very infertile ground. It is interesting to note that Avicenna from Iran developed what would today be considered an approach to experimental design.

Scholars consider medieval philosophy to have dealt with three problems: the eternity of the world, divine omniscience and the possibility of human freedom, and the immortality of the noncorporeal soul. As discussed earlier, Aristotle's view that the cosmos had no beginning was closely linked to his abhorrence of emptiness because every change required a preceding change, a belief that became the foundation of natural science in this era. There was no place for emptiness nor the void, thus no need for a digit such as zero, or for that matter even one. However, the revealed religions of Christians, Jews, and Muslims required a beginning of the universe created by the eternal God, where there wasn't much room for emptiness.

The end of the medieval period is generally agreed to have occurred around 1700 CE due to the emergence and ultimate dominance of what some historians refer to as the *Renaissance*, which began as early as 1400. This era followed by the *Enlightenment* can best be described as a rebellion against Christian theology, and thus the philosophy of Aristotle, which led to the Reformation and contributed to the Protestant schism in Christianity. For the history of intellectual concepts, there is no sharp demarcation between these ages. There is instead a constant bubbling of different ideas in most of the intellectual pursuits of scholars in these eras. Part of this rebirth was due to the wider availability of ancient texts coupled with the increased interaction with other cultures. Gutenberg's invention of the printing press in 1448 broadly disseminated the written word, often in local languages. Popular political and philosophical letters and pamphlets were published and social gatherings debating these topics widely spread *Enlightenment* concepts. These texts greatly expanded thought and initiated the contemplation of concepts originally proscribed by Christian doctrine. However, the Church still had a strong hold on scholarly pursuits, requiring many concepts to hone to the official doctrine otherwise the author would face persecution or worse.

Recently, there has been debate as to the exact definition and duration of these historical periods depending on one's perspective and purpose in studying them. The increased access to Arabic texts and the better study of how Christian and Islamic cultures interacted and exchanged ideas has blurred the transition points between eras. For our purposes, by the end of the Renaissance and during the Enlightenment, Christianity was losing its hold on philosophy, the *Scientific Revolution* was fermenting, and society in general was undergoing seismic changes reflected in larger scale warfare. All these disruptive events allowed for the growth of alternative viewpoints including the development of concepts such as the void and zero.

It was during this fulminating period that the concept of the void in a more positive context arose through the reemergence of Lucretius's lost poem *De Rerum Natura*. This projected Epicurus' philosophy into Renaissance scholarly circles including his perspectives on atomism, nothingness, the void, and his empirical philosophical method of reasoning based on senses and testing assumption. The question constantly arises as to why Lucretius' poem, the gateway to Epicurean philosophy and science, was lost for a millennium. A similar question arises as to what happened to the knowledge gained from encountering Indian philosophy when Democritus travelled to India or Alexander the Great invaded it. This knowledge may have existed in various translations buried in monasteries through the Dark and Middle Ages but only emerged when texts from Persia and the Middle East penetrated European institutions. Many scholars believed that since these teachings were so antithetical to the dominant philosophies of Aristotle and tenets of Catholicism, copies that did emerge were most likely deliberately destroyed.

This transition period between medieval and modern times is fascinating and changes in philosophy, religion, and science go far beyond the topics pertinent to this book. Astronomers including Tycho Brahe, Kepler, Copernicus and Galileo, led to the displacement of Earth as the center of the solar system destroying the heliocentric cosmology that dominated thought for millennia. This ultimately allowed the blossoming of mathematical concepts that needed a digit, such as zero, to be fully understood; culminating in the development in the West of calculus by Isaac Newton and Gottfried Leibniz in the mid-1660s.

The seventeenth century French mathematician, natural scientist, and philosopher *Renee Descartes* (1596–1650) (named *Renatus Cartesius* in Latin) plays several crucial roles involving both the void and the number zero. Some modern scholars define the work of Descartes as the transition point between Renaissance and early Modern Philosophy. Although Descartes was trained in Aristotelian philosophy at the new Jesuit College of La Flèche, he was also exposed to Cicero, the Stoics, and importantly the atomists including Democritus. Galileo's discovery of the moons of Jupiter was celebrated at La Flèche in 1610 when Descartes was there. To Descartes, the atom was the basic building block of matter and possessed physical size, shape, motion, and weight. He saw the natural world as consisting of matter possessing *extension* described by a few fundamental properties and obeying a few universal laws when interacting or moving. In Cartesian physics, motion determines the size and shape of individual bodies (Fig. 2.3).

Fig. 2.3 Renee Descartes

In his cosmology, the full universe consisted of the plenum where emptiness was filled with particles; there was no such thing as actually empty space or a void. A void *is* a substantial body in itself because it possesses *extension* which is the essence of a substance according to Descartes. Extension is enough to constitute a body, so voids are substantial bodies. His vortical theory has the entire universe being filled with elements of different sizes constantly shifting around one another. This concept of a substantive plenum gradually withered away in the face of nineteenth century scientists, including Boltzman, Carnot, Joulle and Maxwell, who developed the theory of thermodynamics. This theory is based on the concept that heat was essentially the movement of particles in space or other mediums, and those particles were governed by the same physical laws derived by Newton and the developing physics of the atomic theory. For a book on zero, it's interesting to note that the laws of thermodynamics are listed starting with zero!

This natural world also included an immaterial mind (the soul) that was directly related to the brain, thereby formulating the modern concept of the mind-body problem, a major focus of philosophical discourses on Descartes. Finally, Descartes deserves mention in the mathematical concept of zero as he developed the X –Y perpendicular axes of our modern Cartesian coordinate system pictured below. For the purposes of this book, the origin of this system is the point (0,0). His work influenced Isaac Newton's concepts of motion and provided him with a coordinate system upon which calculus could grow (Fig. 2.4).

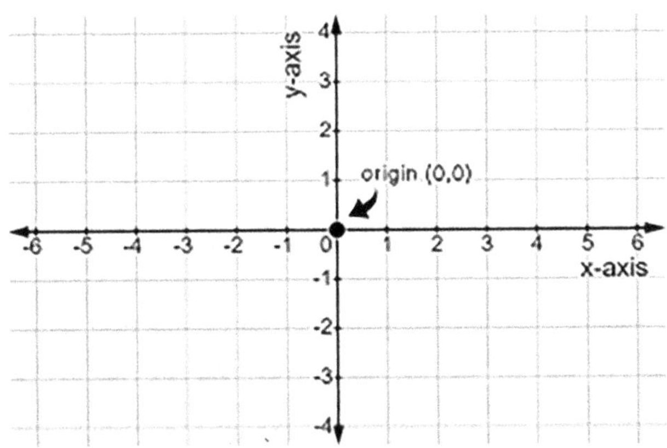

Fig. 2.4 Cartesian coordinate system whose center is (0,0)

This is a good point to illustrate another interesting historical anomaly that I never realized until I was writing this book. As mentioned in the introduction and as can be seen in the above Cartesian plot, zero is the marker or divider between positive and negative numbers. However, when we look at the standard *Gregorian* calendar adopted in 1582 CE and still in use today, there is no year zero between the

BCE and CE timeline. Thus, if counting forward from 5 BCE, one would see the years:

←----● -5 BCE, - 4 BCE, - 3 BCE, - 2 BCE, - 1 BCE, 1 CE, 2CE, 3 CE, 4 CE, 5CE ●----→

There is no 0 BCE or CE! A person born in 5 BCE would only be 9 years old in 5 CE. This is different than the Cartesian number line. It is as if the Cartesian plot above intersected at (1,0) or (1,1) rather than (0,0). Specifically, the names given to the timeline reflects the absolute dominance of the Catholic church on society in general, especially as the dividing event is the birth of Jesus; first conceived by the English monk *Bede* in 731 CE, and then supported by Emperor *Charlemagne* using the date "estimated" for Christ's birth by the Eastern Roman monk *Dionysus Exiguus* in 525. In the 1980s, BC (Before Christ) was replaced by BCE (Before Current Era) and AD (Anno Domini—Latin for in the Year of the Lord and coined by Dionysus) was replaced by CE (Current Era).

As we will see in the next chapter, the digit zero becomes important in our base ten Hindu-Arabic number system as it allows the determination of place and thus magnitude of the digits preceding it (10, 100, 1000, *etc.*). Its psychological effects persist when a new century changes or especially the fervor associated with the dawn of the current millennium (1999–2000; millennial fever struck). The mystery and power of zero strikes again!

Modern Philosophy and the Concept of Emptiness

There are several philosophic traditions that comprise Modern Philosophy, including analytic or linguistic philosophy, skepticism, post-structuralism, and existentialism.

Analytic or linguistic philosophy developed in the twentieth century into what many consider a dominant philosophy today. It emphasizes the study of language and the logical analysis of concepts. It addresses philosophical issues by approaching and analyzing the terms in which they are expressed. Although not focused on concepts like nothingness or emptiness, which are the purview and focus of the nihilistic philosophies to be discussed next, the linguistic approach does raise an issue that is central to this book. In following the arguments of *Ludwig Wittgenstein* (1889–1951) about how language defines the range of thoughts possible to express using it, he has stated *the limits of my language are the limits of my world.* Could zero not have arisen in the West because there were neither sufficient nor appropriate words to communicate the concepts of emptiness, the true void, and what nothing really meant, outside of a strong association with negative and feared emotions?

This contrasts with Eastern philosophy which had a rich history, and thus vocabulary, to debate these concepts that were central tenets of their philosophy. The ancient Indian schools of philosophy and spirituality, including Buddhism, are still studied and remain the basis of Eastern thought up to the present day. The *Dalia Lama* expounds on *emptiness as being the way things really are;* the pronouncement of Nagarjuna discussed earlier from the second century CE that he believes still holds today. In fact, he sees the philosophy of emptiness, as well as many tenets of some of the ancient Buddhist schools, well at home with the latest constructs of modern quantum physics. He and other Eastern philosophers realize that certain tenets, including the philosophy of *emptiness*, may apply in the search for the *ultimate* truth, but do not hold much to explain the everyday existence of *conventional* life where cause and effect is a reality; beliefs very similar to when quantum *versus* traditional physics are needed to explore the natural macroscopic domain. The seventh century CE Indian scholar *Chandrakirti* held that when one formulated an understanding of reality, one must be aware of the scope and specific focus of the inquiry. As will be seen when we discuss the interpretation of what a measured zero today means in certain contexts, this flexibility in language that allows interpreting a phenomenon relative to the domain in which it is being measured or explored, is crucial to interpreting a measured zero across different domains.

Philosophical *skepticism* remains a philosophy that challenges the validity of certain knowledge. They range from those who deny all possibilities of knowledge to those who suspend judgment when there is a lack of evidence. Many of these tenets are based on the skeptics we encountered in ancient Greece and Rome. Finally, *post-structuralism* is a philosophy that also focuses on the radical uncertainty of knowledge. Individual values and meaning arise from the common meanings of a social group making their reality contextualized and social construct communicated by language and discourse. Because of its social nature, knowledge is fluid and changes with one's cultural, economic, social, or political position in the world. Proponents challenge accepted facts or beliefs as being dependent upon the power of the actors holding them. They are often used as approaches and tools of textual analysis and literary criticism. These philosophies focus on the basis and foundation of knowledge itself and do not concentrate on the concepts that are the subject of this book.

The concepts of emptiness in modern philosophy leads to *nihilism* and forms the basis for *existentialism*, the philosophy briefly introduced here due to its focus on emptiness and nothing. Existentialism is the key philosophy of the nineteenth and twentieth centuries, whose major proponents among others include Albert Camus, Fyodor Dostoevsky, Martin Heidegger, Franz Kafka, Soren Kierkegard, Friedrich Nietzsche, and Jean Paul Sartre. It is a movement that in the broadest terms addresses one's sense of alienation and despair and forces the ultimate acceptance of man's fundamental aloneness in an indifferent universe. Existential tenets have had major influences on the development of concepts, including anxiety, angst, and anguish in modern psychology. Each individual is solely responsible for both creating purpose or meaning in their lives, but also to a nihilistic philosopher, such as Nietzsche, it is the individual that also provides structure and order to the world. This is very different than the social constructs in post-structuralism. To Heidegger, our existence is what occurs between birth and death. Individuals are the shepherds, not the creators nor engineers of their own identity. Existentialism promotes the concept that one must view and understand life and existence from the inside; it is not forced upon us from an outside power. Our life is the only immutable thing about us, and we remain in a constant process of self-realization. In contrast to earlier philosophies, we have freedom and thus free will. The concept of nothingness pervades many existential writings. *Sartre* opined that every thought requires one to think about something. One cannot simply think, one must think about something, thus there cannot be intuition of nothingness. The core of the modern nihilistic philosophy is simply we are nothing, we are born, we alone create what our existence is like based on what we inherited and what our environment is like. We adapt, we die, and then we are nothing again.

Some view the recent discoveries by the Hubble and Webb space telescopes that show the unfathomable immensity yet apparent emptiness of the universe as underscoring both the insignificance of humans and thus our own planet Earth in the context of the infinite cosmos. Despite the countless numbers of stars and galaxies seen in all of these images, paradoxically space remains largely empty. Despite searching and listening for evidence of intelligent alien civilizations, we have found none. This could be due to either the immense distances existing between potentially inhabitable stars or vast differences in time when these civilizations could rise and fall providing no window of overlap for communication. Contact between civilizations is essentially impossible according to the physics we know today. In addition, some theorists, including Ward and Brownlee, in what they refer to as the *Rare Earth* hypothesis, discount us ever finding any complex intelligent alien life due to the unique circumstances of our planet.

- Earth's local G*oldilocks* location in our solar system and galaxy (presence of large moon, stable sun, orbit allowing liquid water, right number and balance of large planets, right location in a stable and peripheral part of a galaxy's quiet and boring arm).
- A changing yet always inhabitable global environment favoring evolution (tides, plate tectonics, iron planetary core producing a protective magnetic field, periodic periods of hot and freezing epochs).
- An appropriate number of well-spaced cataclysmic events (comet and asteroid impacts), all that promote new species evolution but not global annihilation of all life.

These unique and random events were among many that allowed for and promoted the evolution of humans on our unique planet. Astronomers have formulated various equations to calculate the probability of intelligent civilizations existing today, the most familiar one being the *Drake Equation* originally formulated in 1961 and modified many times since. The conclusion could be that there may be *zero* advanced alien civilizations like ours and thus maybe *we are truly alone*.

Summary

The purpose of this brief overview of philosophy centering on the development of the concepts of the void, nothingness, and emptiness was to give insight into the unique hold these concepts had on philosophers throughout the ages. This philosophical background also influenced those scholars who ultimately developed mathematics from which zero emerged. As will be seen in the next chapter, there is little doubt in my mind that the digit 0 originated from the necessity of developing a counting system to be used in book-keeping, keeping track of time, distances in geography and astronomy. However, the philosophical and Christian aversion to the concept of nothingness may have been influential in delaying its emergence and acceptance in certain cultures.

Nothingness and emptiness were and remain central constructs of major schools of both Eastern and Western philosophies and religions. Psychologically they can be the drivers of both anxiety and depression or the enlightened goals of a meditative *Nirvana*. These discordant perspectives, which we now have a better grasp on as we are aware of their origin story, still affect the interpretation of using the digit 0 today. When interpreting 0 as *nothing* or *empty*, all sorts of contexts derived from numerous encounters subconsciously infuse our cognitive processes affecting our perspective on the issue at hand and influence how we come to a conclusion. That negative meme suggesting cautiousness still hangs on. The linguistic limitations described by Wittgenstein on our cognition influence how an individual interprets 0 to this day. *The major issue that this book addresses is understanding how zero is to be measured, interpreted, and employed relative to the problem one is working on.* This sorting process requires great discipline and effort to parse out the different baggage (vocabulary, definition, methods, sensitivity, rules, purpose) associated with *zero* or *nothing* in the different disciplines involved in the analysis, get the contexts involved straight, and achieve a rational and deliberative solution. This in fact is the true goal of philosophic reflection.

Selected Readings

Bottone C. *Quod Inane Vocamus: Lucretius' Void in Seventeenth Century Italy*. New York, *CUNY Academic Works*, 2021.

Cooper D.E. *Existentialism*. Cambridge, Blackwell, 1993.

Dalai Lama. *The Universe in a Single Atom: The Convergence of Science and Spirituality*. New York, Morgan Road Books, 2005.

De Harven V. How *Nothing Can Be Something*: The Stoic Theory of Void. *Ancient Philosophy* 35:405-429, 2015.

Fernandez-Armesto F. *Out of Our Minds: What We Think and How We Came to Think It*. Oakland, University of California Press, 2019.

Hatfield G. *Rene Descartes. The Stanford Encyclopedia of Philosophy*, Zalta EN (Ed). https://plato.stanford.edu/archives/sum2018/entries/decartes,2018.

Fan M, and Sullivan IM. The Significance of Xuwu (Nothingness) in Chinese Aesthetics. *Frontiers of Philosophy in China*, 5:560–574, 2010.
Kaplan R. *The Nothing That Is: A Natural History of Zero*. Oxford, Oxford University Press, 1999.
Mensch P (Translator), Miller J (Editor). *Laertis Diogenes—Lives of the Eminent Philosophers*. New York, Oxford University Press, 2018.
Marenbon J. Medieval Philosophy, *The Stanford Encyclopedia of Philosophy*, Zalta EN , Nodelman U (eds.), https://plato.stanford.edu/archives/spr2023/entries/medieval-philosophy/. 2023.
Marino G. *The Existentialists Survival Guide*. New York, Harper Collins, 2018
Marino G. *Basic Writings of Existentialism*. New York, Modern Library, 2004.
McFadden. *Life is Simple: How Occam's Razor Set Science Free and Shapes the Universe*. New York, Basic Books, 2021.
Passannante G. *The Lucretian Renaissance : Philogy and the Afterlife of Tradition*. Chicago, The University of Chicago Press, 2011.
Reeve CDC. *Plato Republic*. Indianapolis, Hackett Publishing, 2004.
Robertson R. *The Enlightenment: The Pursuit of Happiness: 1680-1790*. New York, Harper, 2021.
Sartre JP. (Translated by HE Barnes). *Being and Nothingness: A Phenomenological Essay on Ontology*. New York, Washington Square Press, 1984.
Seife C. *Zero—The Biography of a Dangerous Idea*. New York, Penguin Books, 2000.
Smith NK. *Descartes Philosophical Writings*. New York, The Modern Library, 1958.
Sorensen R. *Nothing—A Philosophical History*. Oxford, Oxford University Press, 2022.
Ward P and Brownlee D. *Rare Earth—Why Complex Life is Uncommon in the Universe*. Göttingen, Copernicus Publishers, 2000.
Wanjek, C. *The Drake Equation*. Cambridge, Cambridge University Press, 2015.
Zalta EN, Nodelman U. Plato, *Stanford Encyclopedia of Philosophy*. https://plato.stanford.edu/info.html#c, 2023.

Chapter 3
History of Zero as a Number

Numbers rule the universe. (Pythagoras)

Any consideration of the history of zero must first discuss the origin of numbers, specifically the counting integers. *Central to the focus of this book, zero doesn't appear as an actual number in any ancient numbering system!* As we can now appreciate, the emergence of zero as a number in itself, and not in its role as a position marker, may have been delayed due to its cognitive association with the void, which had such strong negative connotations that a number representing nothing was unthinkable. Philosophers concerned with such an issue were often the same ancient scholars who were also astronomers and mathematicians who would have benefitted from its use in computations. This delay is unfortunate because zero's final emergence only centuries ago enabled mathematics to blossom and flourish.

There is extensive literature on the intricacies of ancient mathematics, as many of their constructs still underpin modern practices, especially in algebra and geometry. Our goal will be to briefly overview the growth of ancient number systems, introduce the concept of zero in counting and its use in positional numbering systems, and discuss the emergence and ultimate dominance of the zero-containing Hindu-Arabic numerals that spread globally to become the basis of our modern numbering system. For those interested in the complete and much more detailed story of numbers, Georges Ifrah's excellent book *The Universal History of Numbers,* or the more recent *The Secret Lives of Numbers* by Kitagawa and Revell should be consulted. As a final caveat, for our purposes in discussing the history of zero, we will only consider the whole numbers that arise from counting, a subset of all natural numbers.

Figure Use Attributions
Figs. 3.2, 3.3, 3.4, 3.6, 3.7, 3.8, 3.9, and 3.10: Wikimedia – No restrictions
All other figures are either self-drawn, photographed or from author's own publications (attributed)

J. E. Riviere, *Zero – Much to Do About Nothing?*,
https://doi.org/10.1007/978-3-031-82998-7_3

Counting

Some 40,000 years ago, numbers based on counting digits on one's hands and feet allowed systems of base 5, 10, and 20 to evolve. In mathematics, *base* refers to the number of different and unique characters used by the numbering system. The binary system is base two and has two characters: 0 and 1. Our decimal system is base ten and has ten digits. One can consider the hand as the first human counting machine and can appreciate where the word *digits,* meaning counting numbers, arose! There is evidence of such systems being developed across many ancient civilizations.

The oldest mathematical artifact, the *Lebombo Baboon* fibula bone from South Africa, was related to counting. It originated around 35,000–40,000 BCE and contained 29 notches suggesting a *lunar calendar*, which some speculate could have also been used by African women to mark menstrual cycles. It resembles the counting sticks still used by Nambian bushmen today. A similar but later artifact, the *Ishango* bones recovered from Congo-Uganda and made in 20,000 BCE, has three columns with 60 notches, interestingly with one section potentially being prime numbers. As one can appreciate, any further speculation on what these prehistoric artifacts were actually made for is not warranted.

The use of digits in counting explains why even early systems without a zero were still base 10 since counting 10 fingers was a universal idea. However, digit counting has limited capacity, no endurance and leaves no archeological record except for few hand stencils in prehistoric French cave paintings. What does endure are *tally marks* etched in stone, wood, or bone as well as etched clay tokens and tablets that were used for counting time or keeping records of quantities. In a tally mark system, 12 would be recorded as "~~IIII~~ ~~IIII~~ II". I still use this system today, often referred to as the *five-barred gate*, when keeping track of game scores using pencil on paper. Some scholars and cognitive neuroscientists believe that the human mind can only keep track of four objects without having to count them, thus when notches were used, the base unit was IIII before striking through them ~~IIII~~ to represent five. This system also required a significant cognitive development in human brains that allowed for a *one-to-one correspondence* or *mapping* between real objects being counted (animals or stones or trees) and their *abstract* representation in the corresponding number of fingers, tally marks, or beads. This evolutionary leap was a significant event in the development of human civilization.

Since the focus of this book is on the origin of zero and not a history of mathematics, we will jump forward from prehistoric civilizations to the merely ancient ones, which developed more sophisticated numbering systems that ultimately led to the need for a digit, such as zero to represent the empty set and for the use of a place setting digit in so-called positional numbering systems, to express orders of magnitude. In a positional number system, a digit's value is dependent upon where it is located within a number, thus the digit **2** in 1**2** *versus* in **2**13 carries very different meanings (values of 2 *versus* 200). It is interesting to note that positional numbering systems were independently invented only four times in human history: Babylon in 2000 BCE,

China around the beginning of the Common Era, by Mayan astronomers around 300 to 500 CE, and rediscovered in India in the fifth century CE. Of these four numbering systems, some form of zero ultimately emerged only in Babylon, India, and Maya.

Cultures that developed sophisticated numbering systems did so for counting objects used in trade or assessing inventories. These would be called *cardinal* numbers because they refer to counting the number of objects. In contrast, *ordinal* numbers keep track of order or sequence or the progression of time as in astronomy.

My coverage of numbering systems by choice is not comprehensive, as other texts on the history of mathematics are readily available. Instead, the focus is on showing where the need for the naught digit zero arose and significantly influenced the development of higher mathematics, which led to its use in a myriad of practical and scientific applications.

Ancient Numbering Systems

Egypt developed the earliest numbering system based on Egyptian *hieroglyphics* dating back to around 3000 BCE. It was a base ten system but did not include zero as a digit nor as a marker in a positional system. Instead, it utilized *pictographs* to represent orders of magnitude ten (Fig. 3.1).

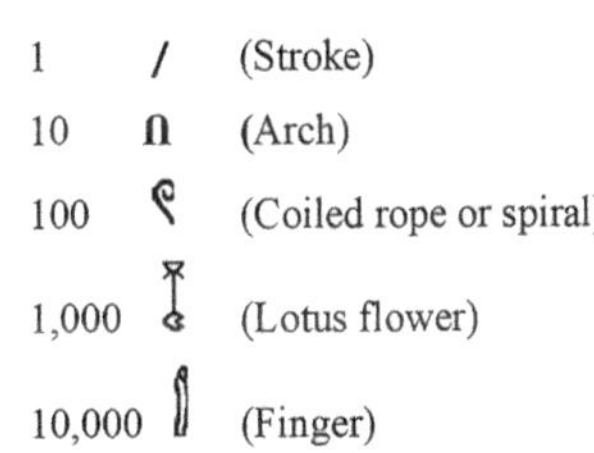

Fig. 3.1 Egyptian numeral pictographs

Since it was not a positional system, the symbols held their value regardless of location. The number 122 could be written as "𓍢𓎆𓎆//" or "//𓍢𓎆𓎆". The base ten structure of the Egyptian numbering system was also clearly reflected in their spoken numbers that have words for the digits 1 through 9, then the decades 10 through 90, and finally for 100, 1000, and 10,000 upwards. A major advantage of this system was that arithmetic, especially addition and subtraction, was relatively easy to perform. Symbols were added, then regrouped into the next highest order (////////// in a sum would be replaced by 𓎆). Multiplication by ten was also easy because it only involved substitution for the next higher order of magnitude. Some workers credit the *Rhind papyrus* written by the Egyptian scribe *Ahmes* in 1550 BCE (but probably collated from texts from as far back as 2000 BCE) as representing the first mathematical textbook ever written. It included many principles of trigonometry, which obviously is not surprising because the pyramids were constructed using its tenets. One can appreciate that even in this ancient civilization, the rudiments of a numbering system facilitating arithmatic manipulations existed. However, there was no zero.

The *Babylonians*, who were well known and proficient in astronomical observations and for the invention of the abacus, used a *sexagesimal* (base-60) numbering scheme based on earlier *Sumerian* and then *Akkadian* civilizations (approximately 2500 BCE), the former being the true originators of *cuneiform* (wedge shaped symbols) counting on clay tablets. These are the only civilizations that developed a sexagesimal system. Some mathematicians have speculated that 60 was convenient as it is the lowest number with the greatest number of divisors [1, 2, 3, 4, 5, 6, 10, 12, 15, 20, 30, and 60], the lowest number with the first six integers as factors, and the number that worked well in astrometrics. Others speculate that exposure to a base 12 *duodecimal* system, originating from fusion with earlier civilizations used to count and measure objects [dozen, gross (12 × 12), 12 inches in a foot, *etc.*], combined with a base 10 system resulted in 60 being adopted as the lowest common multiple of 10 and 12 (6 × 10 and 5 × 12). Obviously, the ancient inventors of this system had no knowledge of these properties, but rather these properties may have allowed a base 60 system to be more versatile and "easy to use" in the natural world from which they arose. Babylonians often did not use any symbol for 60. Some also postulate a mystical connection since the Akkadians had infused number symbolism into Babylonian thought. The full circle with 360° (degrees) represents the totality of the cosmos while 60 was the figure of the absolute God *Anu*.

This *sexagesimal* system remains in use today as our basis of telling time, where 60 s are in a minute and 60 min in an hour. It also remains integral to geometry and astronomy where a circle is divided into 360°, with each degree being 60′(arc minutes) and each arc minute being 60″ (arc seconds). Remarkably, this ancient base 60 numbering system, developed some 4000 years ago, coexists today with our Hindu-Arabic base 10 decimal numbers. The origin of using base 60 is not as direct as using hands and feet for base 10 or 20. Over time, base 10 subunits were employed to facilitate tabulation and around 1500 BCE, gradually became the dominant system overtaking the sexagesimal system for most everyday counting and arithmetic. The sexagesimal system persisted for other scholarly uses such as in astronomy. As I was writing this chapter, the new discovery of a 3700-year-old Babylonian tablet (*Plimpton 322*) was reported, which suggested that the ancient Babylonians were doing trigonometry with ratios. Thus, both the Egyptians and Babylonians were using these principles 1500 years before the Greeks who were often credited with their invention. We will discuss more below on ratios and their relation to having a number such as zero.

A symbolic base 10 system used cuneiform marks for the digit 1 (vertical wedge 𒁹) and 10 (chevron 𒌋). These marks are easy to create using a stylus on a clay tablet. Interestingly, the symbol for 60, when it was present, was just a larger vertical wedge. Other symbols were used as multipliers to represent larger orders of magnitude (*e.g.,* 100 or 1000). All numbers could be easily written by combining, clustering, or clumping these symbols and using the multipliers to represent larger numbers. This system evolved in the hands of Babylonian scholars into a positional system where place now represented value. Finally, by using such a base 10 system, Babylonians are also credited with inventing the abacus, the ideal computing machine for this system as columns representing place values could be occupied by beads and arithmetic performed by moving them (Fig. 3.2).

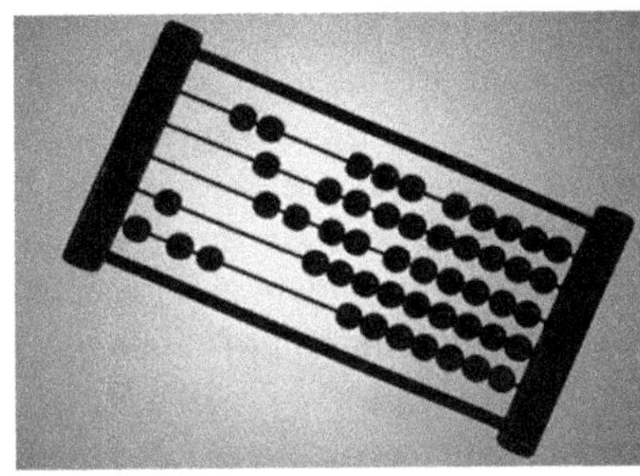

Fig. 3.2 Abacus

It is at this point in the development of numbering systems that we finally get a hint as to where the digit zero may first have originated. The recording of cuneiform digits on clay tablets with a stylus allowed large clay tablet ledgers to be constructed for basic record-keeping functions of accounting (Fig. 3.3).

Fig. 3.3 Cuneiform tablets

This positional system was hampered by not having a special sign to represent units missing from the number to be presented. For example, in writing 102, one can write the symbol for 100, leave a space and then the symbol for 2. However, is there really a space present to represent an absent digit or is it just sloppy scribing? It is estimated that at around 600–300 BCE, two millennia after the origin of this numbering system, Babylonian scribes and astronomers began using special signs to mark the absence of units of a specific order of magnitude. These utilized double oblique wedges () or sometimes superimposed chevrons, symbols that were also used as punctuation in writing. In their use in counting, they now had the meaning of a *void or empty space on a clay tablet* or *nothing in a column of an abacus*. It simply denoted "*nothing in this space or column*" and was not a true number meaning no value or none. Importantly, such symbols were only placed in internal positions within a number and never as the leading or trailing digit. *They were clearly place markers and not ordinal counting digits.* In 331 BCE, Alexander the Great invaded Babylon and brought the Babylonian place marker with him back to Greece.

Chronologically, the next numbering systems to be discussed will be that of the Greeks and Romans. Greece was a center of advanced mathematics in the ancient Western world, though scholars have postulated that because the focus of Greek mathematics and cosmology was primarily based on geometry and trigonometry, fields where a zero was not essential, it never emerged except for its later use a place marker. Greek mathematicians, such as Pythagoras, Euclid, and Hipparchus, focusing on the calculation of ratios (*e.g.*, trigonometric ratios of right triangles such as sine and tangent, the Golden Ratio: ≅1.618, the Ratio of Parthenon: 2.25 to 1) did not encounter zero, for we now know that any ratio of **a/b** requires **b** to be a non-zero integer. In geometry zero has no shape and therefore no geometric equivalent. Greek geometry was the basis both for their cosmology as well as their architecture.

The *Greeks* used a base ten system originating around 500 BCE with digits based on their 27-letter alphabet (3 letters in this ancient alphabet are no longer used in the modern alphabet). The first nine digits are represented by the first nine letters, with the next nine letters denoting tens from 10 through 90, and the last nine letters representing hundreds from 100 to 900. Numbers were then represented as combinations of these alphabetic digits. To differentiate these counting letters from ordinary written letters, a stroke was placed above them. This system allowed for some rather complex arithmetical calculations. The Greeks also adopted the sexagesimal numbering system from Babylonian astronomy. Despite this sophistication, a true zero was not in this system.

They named their place maker the *obol* (the smallest value Greek coin) represented by the lower-case Greek letter omicron (**o**). There is some discussion that the symbol originated as a • in writing, which then transformed into an open dot **o**. In Greek mythology, the obol is the fee paid to Charon, the ferryman of Hades, who transported the dead across the rivers Styx and Acheron that separated the land of the living and the dead. A single obol was placed in the mouth of the diseased at burial as sort of a toll for the crossing. This author cannot help to note the association of even the obol place marker with Hades, seemingly a common thread that links zero with negative connotations! This close relationship between letters and numbers supports the argument that linguistic influences on the use of zero in language could influence its interpretation as a number.

In the numbering system, **o** was not a true integer, but simply a punctuation mark for notational purposes. It was often employed in calculating astronomical values with the results reported in their alphabetic number system described above. Such a symbol for zero used as a place marker appeared in *Ptolemy*'s the *Almagest* in 150 CE, which was the earliest work on mathematical astronomy. Scholars thus do not credit the Greek obol as being the precursor of the modern number zero.

The final ancient Western numbers to be discussed are *Roman* numerals (Fig. 3.4).

Fig. 3.4 Arabic and Roman numbers

1	I	11	XI	50	L
2	II	12	XII	100	C
3	III	13	XIII	500	D
4	IV	14	XIV	1000	M
5	V	15	XV		
6	VI	16	XVI		
7	VII	17	XVII		
8	VIII	18	XVIII		
9	IX	19	XIX		
10	X	20	XX		

For example, 122 in Roman numerals would be CXXII. As can be appreciated, this is also a base ten but not a true positional numbering system, though individual numbers like IV and VI could be considered positional. The individual numbers were created by addition or subtraction (8 = VIII while 9 = IX). Compared to more modern Roman numerals still in use today, ancient Romans used IIII rather than IV for 4, XIIII rather than XIV for 14, ↓ for 50 and ↄ|c for 100. As with all other number systems presented, there was no number zero. Even the comedian George Carlin has realized "*How do you write zero in Roman numerals?*"

Modifications over time to the Roman numerals allowed accounting for much larger quantities by using a *vinculum* (horizontal line) over a number, which multiplied it by 1000. Despite such modifications, Roman numerals were difficult to use in complex calculations yet remained the dominant numbering system in the ancient world up until about the twelfth century CE when Hindu-Arabic numbers replaced them. Although Roman numerals are still in use today for purposes such as hours on clock dials, bullet numbering in manuscripts, dates on public buildings, lineages of popes and monarchs, and sporting events such as the American Super Bowl; they are not generally used in daily commerce nor scientific work, although they do appear in some genetic nomenclature.

The next sophisticated numbering system are those of ancient China which originated in primitive forms anywhere from 1600 to 1000 BCE, though some argue that mathematics was central to the Chinese since the first *Xia Dynasty* founded by *Yu the Great* some 4000 years ago. The counting rod system was firmly established by

770–440 BCE and was sophisticated enough for Chinese mathematicians to extract square and cube roots of numbers and calculate the value of π to 8 decimals (3.14159267)! Yet, despite this history and sophistication, again zero was nowhere to be found. The Chinese system employed a decimal number system that also used a place value ordering. Numbers were vertical or horizontal strokes, that reflected a clustering of bamboo computing rods or *suanzi* organized into two distinct but related sets; the *zeng* for even powers of ten (one, tens, hundreds, ten thousand, etc.) and the *hang* for odd powers of ten (thousands, hundred thousand, *etc.*). In this system, 1, 2, 3, 4, 5, 6, 7, 8, 9 were written as (Fig. 3.5):

Fig. 3.5 Chinese numbers

The initial five numerals are simple vertical tally marks, six is then represented by a horizontal mark coupled to a single vertical stroke, seven by two vertical strokes, and so forth. These nine digits were then used to further count to successively larger numbers using horizontal rods beneath the hang tallies to represent tens: 10 would be –,20 = while 11 would be ⊥,12 ⊥| and so forth. In the second century BCE, negative numbers were denoted by red rods with a square cross section and positive ones by black rods with a triangular cross section. Note that there was no placeholder between the positive and negative numbers.

Recent research has shown that the Chinese used a decimal-based place-value system well before the Common Era. This system allowed complex computations to be performed on a grid where columns represented the order of magnitude of the rods within a space, including decimal columns to the right of the unit column. This grid system evolved in the thirteenth century CE Ming Dynasty to the bead abacus or *suanpan*. Multiplication and subtraction were accomplished by repeated addition or subtractions. Some fourteenth century Chinese mathematicians were also competent in some aspects of algebra. This system then spread to Korea and Japan. The symbol O appeared as an empty placeholder in the twelfth century CE but was not present in the ancient systems, which often just left a space between adjacent digits where today a zero would be placed. Chinese computing rods which also could be made from bamboo, bone, ivory, or jade using this vertical/horizontal placement scheme, were continued to be used up through the seventeenth century.

The final ancient numbering system to be briefly discussed is that of the Central American *Maya* civilization originating around 2600 BCE and who are believed to be the earliest adopters of zero. Their base 20 positional number system (*vigesimal*) consists of three symbols; a shell for zero, a dot for one, and a horizontal bar for 5. Numbers were created in a similar fashion to the Chinese system. The sequence 0 through 19 was written as (Fig. 3.6):

Fig. 3.6 Mayan numbers

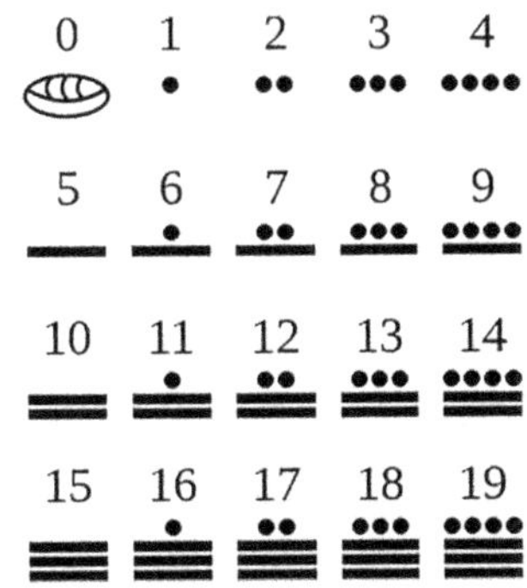

Thus 12 would be represented as two dots over two horizontal rods. Numbers greater than 19 were denoted by adding a multiplier dot atop the number, thus 32 would be the two bars and two dots (representing 12) with one dot above this adding 20. As can be appreciated, addition and subtraction were relatively straightforward. The Mayans had two calendars, the *Long Count* to record events over long periods since the creation of the universe anchored at day zero, and the Civil one (the *Haab*) to record events within a year consisting of eighteen 20 days months with an additional five phantom days squeezed in. The number 360, roughly representing the length of a year was written as one dot over two zeros. In the Haab calendar, zero was used at the beginning of each month with the remaining 19 days using the symbols discussed above. *The earliest record of zero used by the Maya in this system was 36 BCE, a date well before its appearance in Western or Eastern civilizations.* Some scholars hold that Mayan astronomical predictions were more precise than their European contemporaries. For example, the Mayans had calculated a year as 365.2422 days and a month as 29.5302 days, which is essentially identical to the modern estimate of 29.53059. However, by the time Spanish explorers encountered the Mayan culture in the sixteenth century, it was already extinct for generations and their invention of and use of zero had no direct impact on the development of modern mathematics. Some hold that Spanish conquistadors with their Catholic missionaries encountered fragile Mayan texts and burned them as they were believed to be blasphemous.

Exploring these ancient civilizations and their mathematical systems is a fascinating journey into the creativity of human culture throughout the world. One can appreciate the evolution of mathematics and computational strategies and realize the limitations that these cumbersome numbering systems had on doing even simple arithmetic. Similar to the argument of cultural linguists that the language one speaks influences the way one thinks (*Sapir-Whorf hypothesis—linguistic relativism*), the nature of one's language, and thus also their numbering systems, would have a strong influence on what the mind could conceive of using the tools (numbers) at hand (recall Wittgenstein's postulate). In complex numbering systems without zero, many higher levels of mathematical development may just have never been an obvious next step. From our perspective, zero as a number, and not a punctuational place marker, never appeared in any of the ancient numbering systems except for the Maya. Due to its geographical isolation and only late rediscovery after the Hindu-Arabic zero had already penetrated Western Europe, it had no impact on the development of modern mathematics.

Surprisingly, some cultures today still do not have a zero nor even real numbers, relying instead on concepts such as "not much" or "a lot". The *Pirahã* and *Munduruku*, both peoples indigenous to Amazonia, are two cultures lacking words for numbers but retain a sense of scale.

I am fully aware that the evolution of societies over centuries is complex, involves interactions with other cultures, and is often dependent on the perspective and intentions of the historian. Naming ancient cultures across vast swatches of time is also prone to generalizations, errors, and an obvious bias for the historic Western interpretation of ancient texts, as they were incorporated into what ultimately would be modern mathematics. My goal, similar to the last chapter, is to take a broad overview of how concepts, such as the void or zero, developed in these societies. Names applied to these cultures are often prone to biases. For example, as we enter the Common Era, Islamic cultures are described in terms such as Arabian, Babylonian, or Persian by Western historians, yet Jewish contributions were also infused in these traditions. Similar confusion arises when discussing Indian, Hindu, or Jain contributions to this field. In all these cases, the historic record of when zero specifically arose is scarce and in many cases nonexistent. The point of this overview is to specifically track the origin of zero through Eastern, Middle Eastern and Western societies into the Enlightenment and Scientific Revolution; where modern mathematical and scientific practices incorporating zero that define the modern world arose.

The Arrival of the Modern Zero in India and Its Diffusion into the Arab World

The emergence of our modern numbering system, including the use of zero as a true number, can be traced back to *Hindu* mathematicians living from 300 BCE to 600 CE. As we have seen in the previous chapter, Indian philosophy had no apprehension associated with the concept of nothingness and possessed a vocabulary to discuss these issues. At the beginning of this period, a base 10 system was used with symbols for 1 through 9, as well as names for powers of 10. At the end of this period, a symbol for empty or nothing termed *Sanya* in Sanskrit *or Shunya,* written as •, was introduced and ultimately became what we now know as zero.

As Indian mathematicians further developed their art, zero was transformed from a place marker to a true number. Around the fifth century CE (intriguingly concurrent with the fall of the Roman empire), Indian mathematics emerged as a system that reflected the combined influences of Egypt, Babylon, and Greece; as well as a new numbering system in a Babylonian style whose base was ten and not sixty. Our current numbers clearly evolved from these early nine Hindu digits with zero later emerging somewhere in the sixth through ninth centuries.

Zero became its first number with the system now making up the ten digits we work with today; that is 0,1, 2, 3, 4, 5, 6, 7, 8, and 9. Therefore approaching the end of the first millennium, the basis of our modern numbers containing zero was firmly established.

Unlike the Greeks who focused on geometry, Indian mathematicians could use their place number system to do complex calculations without the need for an abacus. Since geometry wasn't their focus, the occurrence of irrational numbers that may occur when calculating a diagonal of a square or manipulating three dimensional solids, never arose. This mindset where numbers were stripped of any physical geometrical manifestation, allowed for their manipulation without requiring a physical reality. This ultimately led to the birth of *algebra*. It also allowed the use of negative numbers because "negative" did not imply the existence of any shape that would be a geometrical absurdity.

This historical process culminated in the work of the Hindu mathematician *Brahmagupta* in the seventh century who developed rules to manipulate numbers and deal with negative numbers (positive times positive equals positive, positive times negative equals negative). Most importantly for our discussion, subtracting the same number from itself equals zero. *Zero was finally a real number* with its own place on the number line dividing positive and negative numbers. In 850 CE, the Indian born Jain mathematician *Mahāvīra* authored the earliest text devoted solely to mathematics: *Ganita-Sara-Sangraha (Compendium on the Essence of Mathematics)*. He clearly articulated over a thousand basic rules and procedures in algebra, importantly including rules for manipulating zero (Fig. 3.7).

Fig. 3.7 Bramagupta

Bhāskara in the twelfth century also attacked the issue of dividing by zero, a problem Brahmagupta attempted but could not solve. He reasoned that one divided by zero was actually infinity. Such an answer could not be further manipulated. Zero and its antithesis infinity were truly strange numbers, but the basis of our modern numbers containing zero was firmly established, as were rules for using them in mathematics. One should note that through these centuries of zero's development, numerous words were used to describe it, but its symbol migrated from • to the **o** we use today.

With the role of zero now firmly established in India, the next step which brought zero into modern times was through the work of Islamic scholars. It is here that zero completely takes root and algebra is firmly established. There is debate on whether we should call our modern number system Arabic or Hindu-Arabic. In the past, they were routinely referred to as Arabic numbers, but recent history underscores the importance of the Indian contribution. By around 700–800 CE, the *Abbāsid* caliphate had translated the work of the Indian and Greek mathematical scholars into Persian and Arabic and wrote them on the newly invented paper. Since Islam is the next stop on our journey to the modern era, we will complete our discussion of its development in ancient Middle East civilizations as that is the point of their diffusion into the Western intellect. An interesting point to note is that we began our discussion of ancient numbers in Egypt and Babylon only to return there now invigorated by the inclusion of zero into their mathematical framework.

The simplest statement of the strength of the Hindu-Arabic numeral system was the fact that it was a base 10 positional number system using zero both as a true value number as well as a placeholder. In this system, the position or place of the digit determines its value. Zero serves both as the first digit and center of the number line as well as the marker that no other counting number appears in this position, thus triggering to advance to the next digit to its left by one number. It represented the movement of a number to its next highest base 10 tier.

← -2 -1 **0** 1 2 3 4 5 6 7 8 9 1**0** 11 12 13 14 15 16 17 18 19 2**0**--3**0** --5**0** -- 1**00** -- 5**00** -- 1**000** →

This system allows all the arithmetic operations of addition, subtraction, multiplication, and division to be easily performed as well as more advanced procedures including taking logarithms that became the backbone of modern mathematics. This would be next to impossible using Roman numerals!

The key figure in the consolidation of Islamic mathematics was the ninth century Persian mathematician *al-Khwarizmi*, a geographer and astronomer from Khwarazm after whom algebra gets its name derived from his book *Kitab al-Jabr*. In addition to his being an astronomer and head of the library of Baghdad's *House of Wisdom* in 820 CE, he is primarily remembered as the Father of Algebra. His accomplishments are too numerous to list here but include introducing the methods of reduction and balancing for solving equations, providing the first detailed description of solving quadratic and linear equations, introduced the use of logarithms and constructed tables of trigonometric values of sine and tangents. Significantly, his book of arithmetic *On the Calculation with Hindu Numerals* was translated into Latin in the twelfth century as *Algorithmo de Numero Indorum.* Together, this body of work gave us the word algorithm, cemented the use of the Indian numerals in mathematics, and propelled the decimal positional numbering system into the Western world. The translation of *Kitab al-Jabr* into *The Compendious Book on Calculation by Completion and Balancing* served as the fundamental mathematical text in Europe until the sixteenth century (Fig. 3.8).

Fig. 3.8 Al Khwarizmi

Unlike Medieval European scholars, Muslim scholars minimized the influence of Aristotle and adopted more Atomists views as they were accepting of the void between particles since it was respectable due to their adoption of zero. At one point, belief in Aristotle was even punishable by death. The Islamic philosopher and physician *Avicenna* (980–1037 CE) discussed concepts of absence and the null, and even developed a method of validation based on experiments. Belief in the void and infinity found a home in Islam as they believed God created the universe out of the void. However, zero as a number did not emerge.

In the interface between Medieval Europe and the Middle East, the Jewish system of mysticism and numerology called the *Kabbalah* arose in eleventh Century Spain. In 1280, the Spanish Kabbalist *Moses de Léon* published the book *Zohar, the Light of the Infinity of God*. With the advent of printing, this mystical philosophy spread throughout Europe. Central to Kabbalah mysticism is the number ten. In this system, the Hebrew alphabet is translated into numbers and then manipulated to gain mystical insights from texts such as the Torah. Its greatest entity is God who cannot be assessed by these numbers and instead is termed *Ein Sof*, without end or infinity. A second idea of the Kabbalah is nothingness, which conceives of a vessel containing nothing. As this Jewish sect spread from the Middle East through Spain, it coupled with the introduction of Islamic acceptance of nothingness to ultimately be noticed by Western scholars facilitating the acceptance of zero in other contexts besides mathematics.

At this point we must now leave the ancient world and move to mid second millennium Europe and trace the reemergence of zero into Western mathematics, essentially intact from the work of Islamic scholars. As we discussed in the last chapter, one must reiterate that the historical record of the exact progression of the Hindi-Arabic numbering system, including zero into Middle Age Europe went through censored religious institutions such as monasteries and universities, leaving large gaps in our knowledge of the precise path and timeline of its adoption. The biggest effect was that in the West, where development of mathematics based on known ancient practices and a functional decimal numbering system was stunted for a thousand years, modern mathematics along with zero flourished in the Eastern cultures of India and the Middle East.

Modern European History of Zero

The pendulum of our discussion on the history of zero now moves back to Western Europe. It is estimated that zero was first introduced into Western culture by *Fibonacci* (*Leonardo Pisano*) around 1200 CE. He was the author of *Liber abaci* (*Book of the Abacus*) in 1202 that brought the work of Indian and Arabic mathematicians as well as the Hindu-Arabic numbering system to Europe. As a youth, his father, a merchant in Pisa, had him spend time in North Africa where he learned arithmetic from an Arab master mathematician, probably using *Algorithmo de Numero Indorum* as a text. He journeyed to Egypt, Syria, Greece, Sicily, and Provence to further study numbers and techniques of calculation. Italian merchants began adopting this system to practical problems including barter, money changing, weights, and measures and calculation of interest. In the Italian double-entry bookkeeping system, zero became the balancing point between negative (debt) and positive (credit) amounts. Once adapted by Italian traders plying the Mediterranean Sea, its use rapidly diffused throughout Europe. In the 1220s, *Holy Roman Emperor Frederick II* invited Fibonacci to Pisa where a scholar presented him three problems. This encounter resulted in a long-standing exchange between Frederick's court and Fibonacci, which further disseminated his influence (Fig. 3.9).

Fig. 3.9 Fibonacci

He is well known today for the *Fibonacci Sequence* where each number is the sum of the two preceding ones (0, 1, 1, 2, 3, 5, 8, 13, 21, 34, 55—∞), which at increasingly large numbers yields precise approximations of the Golden ratio mentioned earlier, defined as $(1 + \sqrt{5})/2 = 1.618034$. This ratio, an irrational number, is approached by the quotient between each successive pair of numbers in the Fibonacci sequence, *e.g.*, 34/21, 55/34 —, with its approximation more accurate at larger numbers. The increased accuracy at successively higher numbers will be seen in the next chapter to nicely illustrate the concept of limits in calculus. Fibonacci made additional contributions to mathematics but for our purposes, he firmly implanted zero as a true number. Its use then diffused further across Renaissance Europe over the ensuing centuries, catalyzed by the use of the printing press, and became integrated into modern mathematics. Recalling from the last chapter, by the seventeenth century Renee Descartes had already anchored his Cartesian coordinate system at (0,0).

Fibonacci is also credited with developing a word for the digit 0, which ultimately becomes our name zero. The Arabic word that he would have encountered in his studies for zero was *sifr,* which meant empty and emptiness. The Arabic *sifr*, still used in Farsi today, ultimately became the English *cipher*, a synonym for zero. Its Latin homophobe was *zephyrus*. Fibonacci used the term *zephirum* for zero which persisted for a few centuries. In the fourteenth century, zephirum became the Italian *zefiro*, whose Venetian contraction is *zero*. The modern zero was first published in an Italian text in Florence in 1491.

An aside into the history of zero is its use in the base 2 *binary* digit system (0,1). The numbers 0–9 written in binary digits are: 0000, 0001, 0010, 0011, 0100, 0101, 0110, 0111, 1000, 1001, and 1010. *Gottfried Wilhelm Leibniz* has long been credited with its discovery in the late seventeenth century, but recent research has placed the true inventors being *Thomas Harriot* and *Juan Caramuel de Lobkowitz* in the early seventeenth century. *Leibnitz* is credited with the independent development of calculus along with *Sir Isaac Newton,* though recently some workers have discovered that fourteenth century Indian mathematician *Mādhava of Sangamagrama* from the Kerala School of Southern India had grasped some of its basic concepts such as the infinite series. There is no strong evidence that either Leibnitz or Newton was aware of this work. We will further discuss zero and its role in calculus in the next chapter.

Although invented sometime in the seventeenth century, the utility and universal adoption of binary digits and its arithmetic rules had to await the development of modern digital computers in the twentieth century to rapidly proliferate. Binary digits are the elemental language of computers. The basic machine language code, the lowest level computer language directly interfacing with the microelectronic components creating *logic gates* in a computer's central processing unit, are coded in binary digits. Materials that can express two physical states (on/off, +/−, conducting electricity / non-conducting electricity, clear/opaque, light pulses, *etc.*) could be used to construct basic computer architecture because they can reflect binary numbers. Electronic digital computers use logic gates, based on resistors or diodes, that are combined to form the integrated circuits seen in all computers today.

Higher level languages, such as Fortran, Basic, Python, C^{++}, or Swift, must then be compiled or translated down to machine code. The tremendous speed and capacity of modern computers is not based on non-binary architecture, but rather on shrinking electronic components that allow more logic gates per chip and thus a shorter path of connectivity that increase capacity and speed. Until quantum computing architecture takes hold, the basic language of all digital computers and devices is the binary code invented in the seventeenth century. As will be seen in the chapter on physics, zero will also play a role in defining the limits of size and speed that material scientists and electrical engineers can achieve.

The use of zero used in mathematics also underwent a broader diffusion throughout European society as its negative meme finally faded. The architect *Filippo Brunelleschi* developed the concept of *depth perception* in architecture in 1425 by projecting on two dimensional pictures images of three-dimensional buildings through focusing the eye to the vanishing *point* in the back of the picture. *Leonardo DaVinci* taught how to use the concept of the vanishing point (•) to provide perspective in paintings, which resulted in much more realistic art (middle frame). The vanishing point, perceived by the eye to be at an infinite distance in the background, is actually a point that has zero dimensions! (Fig. 3.10)

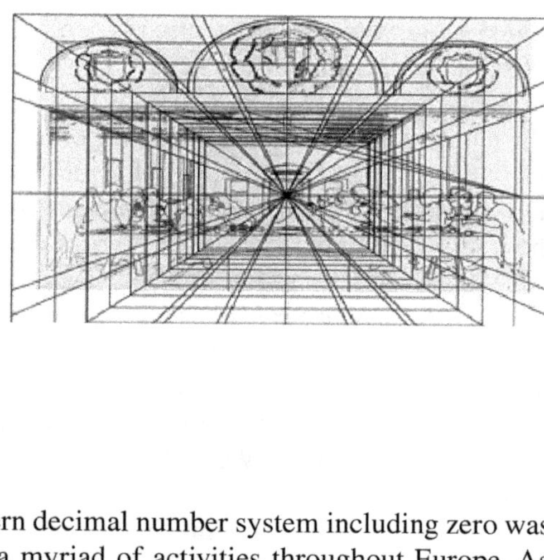

Fig. 3.10 Illustration of the vanishing point used by DaVinci to provide perspective in paintings such as the Last Supper

Summary

By the seventeenth century, our modern decimal number system including zero was firmly established and employed in a myriad of activities throughout Europe. As presented in the last chapter, the seventeenth century coincided with the end of the *Medieval* period and saw the emergence of the *Enlightenment* and the *Scientific Revolution*. Travel and trade increased as did cross-cultural interactions, greatly facilitated by printed pamphlets, journals, and books. Universities independent of clerical control were established. New philosophies emerged and, in some sectors, religious dogmas so antithetical to modern science and mathematics started to weaken. The intellectual mind was transformed on most levels and for many disciplines, facilitating a cognitive outlook that allowed modern mathematics and science to fully develop.

Now that we have some idea of the history of zero both from the perspective of its use in philosophy and mathematics, it is time to shift gears and focus on its use in mathematics and the sciences. We will start with mathematics since its peculiar properties compared to other integers introduces the headaches and dilemmas often associated with its use in the sciences, which will be covered in later chapters.

Selected Readings

Aczel AD. *The Mystery of the Aleph—Mathematics, the Kabbalah and the Search for Infinity*. New York, Washington Square Press, 2000.

Chrisomalis S. *Numerical Notation: A Comparative History*. Cambridge, Cambridge University Press, 2010.

Darling D. *The Universal Book of Mathematics: From Abracadabra to Zeno's Paradoxes*. New York, Wiley, 2004.

Ifrah G. *The Universal History of Numbers: From Prehistory to the Invention of the Computer.* New York, Wiley, 1998.

Katz V. The Mathematics of Egypt, Mesopotamia, China, India, and Islam: A Sourcebook. Princeton, Princeton University Press, 2007.

Kitagawa K and Revell T. *The Secret Lives of Numbers—A Global History of Mathematics & Its Unsing Trailblazers*. London, Viking Press UK, 2023.

Sharifian F. *Cultural Linguistics: Cultural Conceptualizations andLlanguage*. Amsterdam/Philadelphia, John Benjamins Publishing Company, 2017.

Swetz, F. The Evolution of Mathematics in Ancient China. *Mathematics Magazine* 52: 10–19, 1979.

Chapter 4
Zero in Mathematics and Statistics

> Over the years, a number of very smart people have learned the hard way that a long string of impressive numbers multiplied by a single zero always equals zero. (Warren Buffett)

We are now at the point that zero has finally been "discovered", the demons surrounding it have somewhat been placated, and now we can explore the mathematics that make it such a unique number. Unlike earlier chapters, we will not focus so much on origin or history of these concepts, but instead deal with the mathematical properties that make zero such a central part of many disciplines and point out where the rules of using zero in arithmetic cause issues in the real world. This book is not a textbook of mathematics nor statistics. My purpose is to illustrate the different contexts that zero finds itself in, the issues that result, introduce specific definitions and rules, and in later chapters identify methods to deal with zero under specific circumstances.

Zero is both a non-negative and non-positive integer that is the center point of the number line. Zero is characterized as an even number (Fig. 4.1).

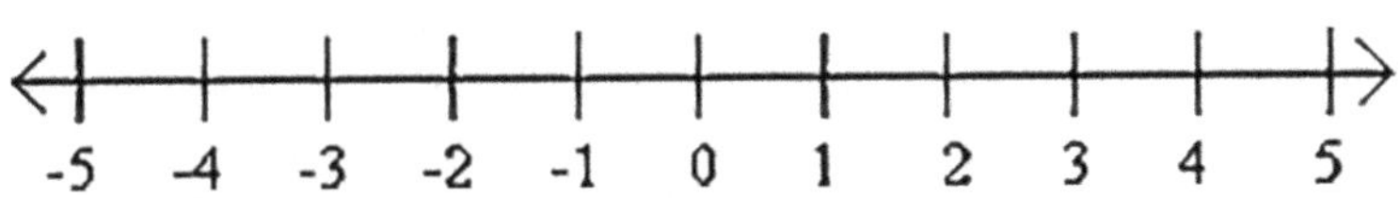

Fig. 4.1 Number line

As an expansion of this one-dimensional number line into the two-dimensional Cartesian coordinate system, the point (0, 0) becomes the origin and intersection point of the horizontal X and vertical Y axes (Fig. 4.2).

Figure Use Attributions
Fig. 4.9: Riviere JE. *Comparative Pharmacokinetics: Principles, Techniques and Applications, 2nd Ed*, Ames, IA, Wiley-Blackwell, 2011. c Jim Riviere
Figs. 4.1, 4.2, 4.4, 4.5, 4.6, 4.7, and 4.8: Wikimedia – No restrictions
All other figures are either self-drawn, photographed or from author's own publications (attributed)

J. E. Riviere, *Zero – Much to Do About Nothing?*,
https://doi.org/10.1007/978-3-031-82998-7_4

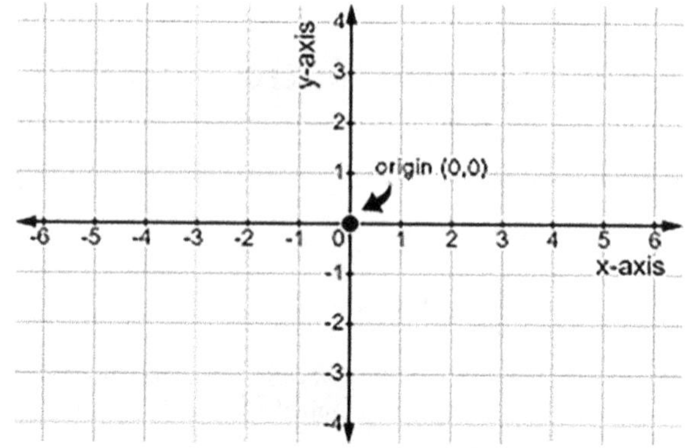

Fig. 4.2 Cartesian x-y graph

We know that the second important function of zero is its use as a placeholder in the base 10 Hindu-Arabic numbering system, defining the order of magnitude of a number. When written as a geometric sequence where each term is a constant ratio with its successor, one gets the sequence:

- -1/1000, 1/100, 1/10, 1, 10, 100, 1000 - - -
which in decimal form is:
- - 0.001, 0.01, 0.1, 1, 10, 100, 1000 - - -.

One can appreciate the simplicity and elegance of this notation, especially compared to the ancient numbering systems we discussed earlier. This property of zero as a placeholder allowed for the development of logarithms, which greatly facilitated manipulations of very large and small numbers.

A *logarithm is the exponent or power to which a base number must be raised to yield a given number*. For our base 10 numbers, if $10^a = b$, then $a = \log_{10} b$. For example, if $10^2 = 100$, then $\log_{10} 100 = 2$. Base 10 logarithms are called common, decimal, or *Briggsian* logarithms and usually are written as just $\log b$. The inverse operation of this allows you to recover the original number *b*. If you know the logarithm of a number is *a*, taking the *antilog* will give you *b*. Today, in most scientific applications, the base *e Naperian* or *natural* logarithm (*ln*) is used (note that in this system the $\ln 10 = 2.303$). *e* is also referred to as *Euler's* number and is an irrational number whose value is 2.718281828459045 ---.

The logarithmic scale was invented by the Scottish mathematician *John Napier* around 1614 and used by the British astronomer *Edmund Gunter* in 1620, which led to the development of the slide rule (Fig. 4.3) by mathematician *William Oughtred* in 1630. Until the development of electric and electronic desk and pocket calculators three centuries later in the 1960s and 1970s, the slide rule was the primary tool used by mathematicians, scientists and engineers to perform complex calculations.

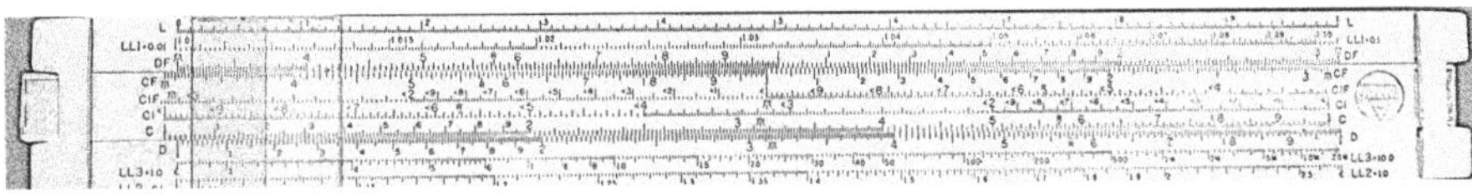

Fig. 4.3 Author's original slide rule

Logarithms turned multiplying large numbers into simple addition problems and division into subtraction. Their discovery and adoption were enabled by the structure of our base 10 numbering system. The use of logarithms to perform calculations with either very small or large numbers facilitated the growth of many scientific applications.

Basic Properties of and Rules for Using Zero in Calculations

To start appreciating why zero has its "issues", Table 4.1 lists the properties and rules for using zero in simple arithmetic. Looking over this, one realizes that zero is indeed a fascinating number! Just consider the manipulations of zero with itself. *Zero either added to, subtracted from, or multiplied by itself gives zero; only dividing by itself is indeterminate.* Euler's identity [$e^{i\pi} + 1$] whose solution is zero has been called by mathematicians the most beautiful theorem in mathematics and the greatest equation ever written. In multiplication, zero becomes the annihilator of all that it is multiplied by. If one can scan an equation and identify that in a product one of the terms will be zero, the equation is solved as its value is zero. Similarly, variables can be eliminated if they can be subtracted from one another or a zero can be found as a multiplier. Balancing equations or their components to equal zero is the fundamental tool in algebra. These properties of zero can both greatly simplify complex calculations or make them indeterminate when zero appears in a denominator or as the operand of a logarithm. In these cases, one could consider zero the *terminator* of further progress, as messages like "Error!" or "Start Again" or "Illegal Operation" will stop the calculation cold. These later issues create most of the headaches with its use. The prohibited operations using zero, dividing by, or taking a logarithm of it, can be studied by calculating functions that approach zero in these circumstances, using the concept of limits, which is one of the foundational principles of calculus.

Table 4.1 Zero in Arithmetic

Addition:
Zero added to itself gives zero. [$0 + 0 = 0$].
Zero added to any number yields the original number. [$a + 0 = a$]
Adding a number to its opposite yields zero. [$-a + a = 0$]
Subtraction:
Zero subtracted from a number equals that number. [$a—0 = a; 0—a = -a$]
Subtracting a number from itself equals zero. [$a-a = 0; -a-(-a) = 0$]
Multiplication:
Zero times any number equals zero. [$a * 0 = 0$]
Division:
Zero divided by any non-zero number is zero. [$0 \div a = 0$]
Dividing by zero is undefined and thus is forbidden. [$a \div 0 = \varnothing$]
If the denominator of a fraction is 0, then the fraction is also undefined.
Exponents:
Anything raised to the power of 0 equals 1. [$a^0 = 1$]
Zero raised to any power is zero. [$0^a = 0$]
Square root of zero is zero. [$0^{1/2} = \sqrt{0} = 0$]
Logarithms:
Log_{10} of zero is not defined.
Log 1 is *0* and thus the antilog of *0* is *1*.
Factorials:
$0! = 1$
Euler's identity
$e^{i\pi} + 1 = 0$ (where *e* is Euler's number and $i = \sqrt{-1}$)

Calculus and the Concept of Limits as Applied to Zero

We now have a good idea on why zero causes difficulties when it appears in certain arithmetic functions. Many of these issues trace back to the ancient Greek philosopher Zeno's paradoxes mentioned in Chapter Two concerning the concepts of instantaneous time or distance intervals. They were also pondered by ancient Indian, Chinese, and Islamic mathematicians. What was needed to solve these issues of instantaneous time, that is a time interval equal to zero, as well as infinitely small intervals as required to solve the slope or area of a complex non-linear function; was the approximate value of a function as it *approaches* the value of zero, not reach it. For example, we all know that to calculate the speed of a moving object, one uses the equation:

$$\text{Speed} = \text{Distance} / \text{Time}$$

How do you calculate the instantaneous speed at an infinitesimally small-time interval approaching zero? We know from the rules of arithmetic that we cannot divide by zero. How do we proceed?

Calculus and the concept of limits provides an answer. It is not the intent of this text to review the development nor practice of calculus, as this is well covered in the textbooks listed in the *Selected Readings* section. Calculus was invented independently in the seventeenth century by the British scientist *Isaac Newton* and the German philosopher *Gottfried Leibniz,* both based upon the work of numerous mathematicians before them. In fact, some scholars believe that its basic principles were articulated 250 years earlier in India though it is hard to know if it impacted later work (Fig. 4.4).

Fig. 4.4 Isaac Newton and Gottfried Leibniz

What is now called modern calculus is based on the work of both Newton and Leibniz, though the name calculus was coined by Leibnitz, who also defined the modern notation for both the integral ($\int$) and the derivative (*dy/dx*). Calculus provides the insight and tools to deal with numbers approaching zero in such a manner that we can solve equations where infinitesimally small values appear in the denominator of a fraction or in other places where a value of zero would result in an indeterminate solution.

A limit is the value a function approaches as the input approaches some value. For our purposes, we want to know what is the value that is approached by the function f(x) = 1/x as x approaches 0.

$$\lim_{x \to 0} \frac{1}{x} = \infty$$

This is easily approximated by calculating the value of $1/x$ for decreasing values x:

x	$1/x$
0.1	10
0.01	100
0.001	1000
0.0001	10,000
0.00001	100,000
0.000001	1000,000
0.0000001	10,000000
0.00000001	100,000,000
0.000000001	1000,000,000

The value of $1/x$ continues to increase as x gets smaller, and thus appears to approach infinity. However, this was the limit when x approaches zero from the positive side of the number line. When x approaches zero from the negative side, this table would be approaching negative infinity, making the function indeterminate. This table nicely illustrates the practical problem of using a target number like a threshold or action limit that keeps getting larger, as seen in analytical chemistry when sensitivity increases, or in toxicology when a threshold decreases. In the scheme above, changing from part per million (ppm: 0.000001) to part per billion (ppb: 0.000000001) dramatically affects the magnitude of a function such as$1/x$! Any function divided by these lower numbers explodes by a factor of ten for each decimal place decrease in concentration.

A zero can also be approached in any function $f(x)$ when x causes the function to approach zero in the denominator. For example, in $f(x) = x^2/(x-6)$; as x approaches 6, the function will approach $x^2/(6-6) = x^2/0$. Logarithms can also cause such issues to arise.

The next place where the concept of zero became important was in the development of *differential calculus* where a *derivative* is the rate of change of a function at a particular point on a function. If this is a simple straight line, it is the slope of the line $(\Delta y/\Delta x) = (y_1-y_0)/(x_1-x_0)$.

If it is a complex non-linear function, geometrically it is the slope of the tangent at a particular point. It is how the function changes with respect to a certain variable, written as $\frac{dy}{dx}$ and solved as a limit of the function's value as dx approaches zero.

$$\frac{dy}{dx} = \lim_{\Delta x \to 0} \frac{\Delta y}{\Delta x}$$

If the *differential equation* is one of a rate process, as we will encounter in chemistry, toxicology, and pharmacokinetic functions in later chapters, the *derivative is dx/dt* where time (t) approaches 0.

The existence of zero as a number allowed such concepts as dt to be conceived of and easily handled by solving differential equations using limits and calculus.

Similarly, determining area under the curves (AUC) of complex functions can be solved using *integration,* which involves calculating limits as some variable approaches a specific value. There are many additional concepts in mathematics that deal with zero, including but not limited to catastrophe theory, singularity theory, the zeta zero of the Rieman zeta function, as well as degeneracy in dynamical systems and bifurcation theory.

Zero in Classification of Rate Equations

Up to this point, we have focused on zero's role in mathematical operations and how limits have enabled it to be used in numerous functions. However, zero also enters the *lexicon of differential rate equations* in chemistry and pharmacokinetics. In this use, it classifies rates of enzyme metabolism or transport in differential equations based on the biological processes dependency on concentration of the compound being acted upon. In this nomenclature, we have zero and first order rate processes:

Zero order: $dx/dt = K_0 * x^0 = K_0 * 1 = K_0$
First order: $dx/dt = K_1 * x^1 = K_1 * x$

The rate is defined by the power to which the concentration is raised. In processes following *zero-order kinetics*, as seen when enzyme systems and transporter pumps have exceeded their capacity and become *saturated*, they show a fixed rate independent of the input concentration. In contrast, in a *first-order process*, the rate is always a function of the concentration available to drive it, a process associated with concentration-dependent *diffusion* in the body. When plotting these equations on the *x-y* Cartesian plot, one gets a straight line for a zero-order process, but a curved line for a first order. When such rate functions are plotted on a Cartesian graph, the curve will asymptotically approach the *y* axis at *0* but never reach it. When the solution for such a differential rate equation for time *t* is instead written in the form $Log X_t = Log X_0 - K * X_t$ and plotted on the *log x–y* coordinate axes (semi-log paper—widely used before calculators were commonplace), a straight line will emerge with slope *K* . Note that because the logarithm of zero doesn't exist, the line extrapolated to zero theoretically never ends. In chemistry and metabolism, there are higher order rate constant equations reflecting increasingly complex processes. Zero used in this context of rate equation has a different meaning from the empty, nothing, or void set and has confused many students and practitioners over the years!

As can be readily appreciated, without the concept inherent to the number zero, calculus would not have developed into the tool that now is so central to numerous scientific applications. With an appreciation of this role, it is time to move on to the next mathematical discipline where zero and the concept of nothing became important. For our purposes later in the book, because science involves measuring things, statistics becomes central to quantitating phenomenon and making decisions about the safety of numerous processes.

Statistics and the Null Hypothesis

The final mathematical discipline where zero has played a role is in statistics, the field that deals with estimating uncertainty and providing methods whereby the potential for chance to cause a specific outcome can be evaluated. As with our discussion of calculus above, this is a vast field with excellent statistics textbooks available for review as listed in *Selected Readings*. Zero and the concept of null plays several roles in statistics, three of which we will briefly be discussed: regression—correlation analysis, confidence intervals, and hypothesis testing.

In estimating the value of the slope of a regression line (an equation relating the value of y for different values of x), the default assumption is that the line has no slope (hence no relationship between x and y) and is thus equal to zero. This is a statement of the *null hypothesis* that there is no slope and thus no relationship between the variables, with *null* as we know being a synonym of *nothing*. In many instances due to uncertainty in a data set, a *confidence interval* is constructed where the point estimate of a number is bracketed by a range of uncertainty, the size of which is a function of the variability underlying data as well as the probability assigned by the user as to how much uncertainty can be tolerated. This is the *margin of error* often encountered in political polls around elections. For example, a survey could estimate a result as $55 \pm 10\%$ voting for a specific candidate. That means the true estimate for our number would be *45–65%*, the latter actually suggesting the candidate could be behind in the race. In pharmacology, a model could estimate drug concentration as 1.0 ± 1.5 mg, the confidence interval is now -0.5 to 2.5, with zero included in the estimate. If this mass were used to normalize a metric of an individual's response to an average dose of drug received, we could have zero appearing in the relationship's denominator, an illegal operation.

There is an area of statistics termed *experimental design* devoted to categorizing the source of uncertainty in estimating real data, including medical or biological phenomenon. Depending on the size of the data set and the design of the study, one can break study units into categories or groups that allow estimation of different sources of variation using a technique termed *Analysis of Variance (ANOVA)*. Errors are estimated for analytical uncertainty, intra- and inter-individual differences, lack of model fit, and random error. The validity of a model is tested by comparing the model variation to the size of the variation that would be expected from all error sources in the overall population. Volumes of literature have been written on experimental design concepts specific to a wide variety of disciplines. My sole purpose of mentioning this here is to stress that when we encounter a zero being measured in subsequent chapters, there is ALWAYS experimental error present and if the assay is not sensitive, the confidence interval may include zero. We will be discussing threshold concentrations defining zero where measurement errors in both the time a sample was taken as well as in the method to analyze the concentration occur. It cannot be stressed enough that in practical applications, zero is defined by the context of the situation or the specific experimental design and it is not a point estimate, but rather one characterized with uncertainties in the x and y axis (Fig. 4.5).

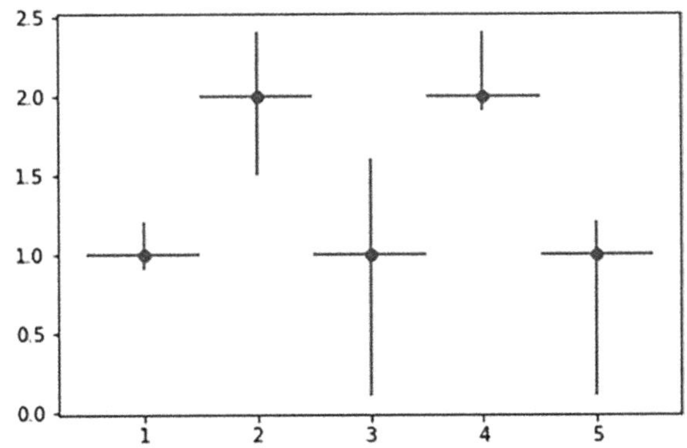

Fig. 4.5 Illustration of data with error bars in x and y axis

Hypothesis Testing is the primary role of statistics in the sciences. As discussed in the Introduction, statistical testing involves estimating the probability that the *null hypothesis* (H_0), generally reflecting what is commonly believed to be correct, the current scientific hypothesis, or the standard model equation, is wrong and should be rejected. This is tested against the *alternate hypothesis* (H_1) reflecting the new data or model that the researcher is proposing. The goal and success for the researcher is to reject the null hypothesis. This alternate hypothesis generally would reflect a new association between the variables being studied or a revised estimate based on new parameters.

Recall that the entire concept of having to prove the null hypothesis is wrong before moving forward is the *Taoist* proposition that *doing nothing may be the best action*, which could be translated to *leave well enough alone unless there is sufficient data to make a change* or the American colloquialism *if it ain't broke, don't fix it*. In medicine, this is consistent with the Hippocratic Oath *primum non nocere - Above all, do no harm.*

In risk assessment practice, this also reflects the European Union's *Precautionary Principle* of risk management where potentially adverse effects should be strongly avoided. The precautionary principle states that when an activity threatens harm to human health or the environment, precautionary measures should be taken when some cause-and-effect relationships are not fully established scientifically. It implies that the introduction of a new product or process whose ultimate effects are disputed or unknown should be resisted until definitive data is available. The basic idea underlying the precautionary principle is often summarized as *better safe than sorry*. This perfectly coincides with the statistical practice of accepting the null hypothesis unless sufficient data is present to reject it for a better alternative.

This is different than the *probabilistic risk assessment* approach of some US regulations where the probability of a beneficial effect is weighed against that of an adverse effect. It was developed in the 1970's to handle the evaluation of risks associated with nuclear power plants. It is applied to complex technological systems and practices, such as aircraft design, spaceflight, toxicology, food, and drug safety as well as evaluating the effects of natural and human induced stressors to the environment. To make a proper decision, sufficient high-quality data in appropriate models are required. This topic will be extensively discussed in later chapters.

Depending on the nature of the data being analyzed, a statistical distribution is usually assumed for the data set (*e.g.*, *normal distribution*) and an estimate of confidence is assigned that reflects how much risk one is willing to accept that the null hypothesis may be true compared to the new hypothesis being tested. This is termed a *Type I error rejecting H_0 when H_0 is actually true* with a probability or *p*-value of α. If α is assigned to be $p \leq 0.05$; meaning there is only a 5% chance or less (1 out of 20) that in fact the null hypothesis was true, then your data set or experimental result did not change the status quo. If α was set at $p \leq 0.01$, then there would only be a 1% chance (1 out of 100) that your hypothesis was wrong. Some workers in specific fields use much lower *p* values to reduce uncertainty. The opposite error is a *Type II error* estimated by β, which is the probability that the *H_0 is accepted when in fact it is false and H_1 is true*. This is related to the power of the experiment with higher power related to larger sample sizes. When making these estimates, the nature of the underlying distribution of the data is critical. In most cases it is assumed that data are normally distributed (*Gaussian* distribution, *Bell Curve* seen below), but there are other distributions that can explain the data and thus affect the selection of the form of the test. The *Central Limit Theorem* (*CLT*) of statistics states that *the distribution of sample means approximates a normal distribution as the sample size gets larger*, regardless of the population's distribution. This is the reason that the normal distribution is so often encountered. Mathematical transformations of the data (*e.g.*, logarithmic) can also be made to obtain a normal distribution for data that would never result in a normal distribution. In some cases, no distribution is assumed, and non-parametric statistical procedures are used (Fig. 4.6).

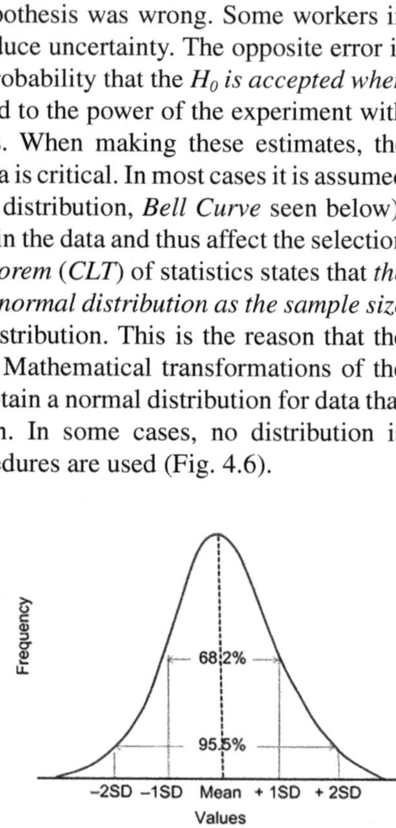

Fig. 4.6 Normal or Gaussian distribution

This plot illustrates the distribution from the mean that a normal population exhibits where 95.5% of all respondents fall within ±2 standard deviations (SD) from the mean estimate. SD is used to estimate confidence intervals. Although the above Bell curve reflects a frequency population analysis, it could represent data plotted from a model with the point estimate now being the mean and the confidence interval expressed as mean ± 2 SD, or if wanting to be even more conservative, mean ± 3 SD, which would cover 99.7% of the data. *This relationship is known as the 68-95-99.7 or Empirical Rule*. Zero could come into play when all of the data falls on the predicted value and the SD is then zero. Statistics is about estimating variability and then making predictions based on how much uncertainty can be tolerated. In toxicology and food safety, that tolerable uncertainty is the risk of adverse health effects occurring.

Another perspective of zero is the concept of the *null set* in mathematics and probability theory. In this approach, there are two kinds of zero, both numerically equal to zero, but with different probabilities. Assume one is choosing a random real number from a uniform distribution for the interval [0,1] containing an uncountable sample space. The probability of getting a number in any nonzero subinterval of [0,1] is finite. For example, the probability of getting a number between 1/4 and 3/4 is 1/2. However, considering the probability of getting an exact number is where differing definitions of zero emerge. The probability of getting 2 is *identically zero* because it's not on the interval and would be called an *impossible* event. However, the probability of getting any exact value between 0 and 1, such as exactly 1/2 is *almost surely* zero. Almost surely has a precise definition of this specific outcome being a zero-probability event, yet paradoxically every point in the interval shares the same almost surely zero probability, yet one event will occur! That is to say 1/2 is a possible outcome but has measure zero. When we consider a countably infinite number of points, the probability of one point is almost surely zero. Such paradoxes abound when one studies very infrequent events in very large samples, an event that often occurs at very small, as well as very large scales. Yet in some cases if the point is outside of the set, the probability is *identically zero.* But if it is just one event in millions, the probability is almost surely zero. There are many cases where this approach can be applied to risk assessment of rare events. When these terms are applied outside the realm of professional statistics, one can see where confusion might arise.

The other topic where zero plays a major role is in analyzing the relationship between two variables, x and y, as presented in a two-way data table. If there is a direct *one to one* relationship between the variables, they will be deemed highly correlated with a *coefficient* of d*etermination* (R^2) approaching 1 or 100% correlation, with all variation explained. R^2 can be thought of as the % of the variation that is explained by the model relationship. Note that some programs (*e.g.*, simple calculators) report R, the *correlation coefficient* as the strength of the relationship between two variables. Even if R is a high value (say 0.8), an R^2 would only be 0.64 meaning only 64% of the relationship between these variables are explained by the regression equation being used and 36% is due to something else.

If one plots this on an x,y plot, then the relationship will be a line with a slope of ($\Delta y/\Delta x$) and an intercept at (0,0). Of course, not all data will appear on this line because of the variability we have been discussing. A line will only pass through the origin (*0,0*) when two variables are related by a simple ratio defining a linear slope. In the case plotted below, the variable x is height and the variable y is weight. R^2 will reflect the amount of data scattered around the line (Fig. 4.7).

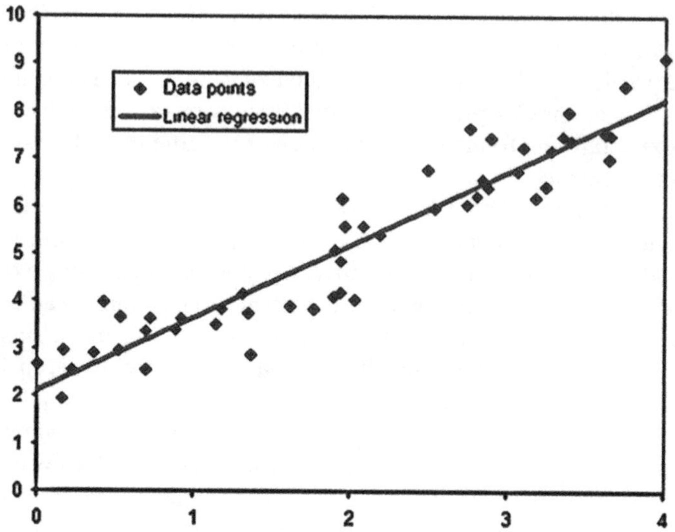

Fig. 4.7 Example of simple regression

In most cases, the line may not begin at the origin and thus the relationship between y and x would be the regression equation of the form $y = mx + b$ where m is now the slope and b is the intercept displaced from the origin. The hypothesis now being tested is the *null hypothesis that m = 0* with the *alternate hypothesis that m is the estimate from the regression equatio*n. This is the form of a calibration curve used in analytical chemistry to validate a chemical assay and predict unknown concentrations where the detector sensing the chemical is correlated to a concentration of the substance. If such an analysis shows no correlation ($R^2 = 0$) and thus a slope of zero (m = 0), there is no validity in the assay. Unfortunately, scientific literature is filled with techniques and assays that really do not have suitable calibration curves, making any estimate of concentrations filled with uncertainty. Often an R vs R^2 is quoted, and then misrepresented as showing say 90% concordance when really the R^2 is only 81%. Many individuals not familiar with chemistry do not realize that the signal being used to detect the presence of a chemical (light absorbance, mass, fluorescence, etc.) is *only valid if* the calibration curve relating the signal to the actual chemical concentration is accurate and robust (Fig. 4.8).

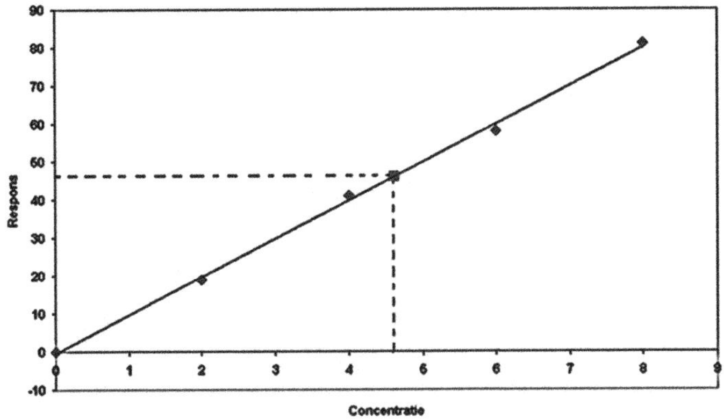

Fig. 4.8 Sample calibration curve as used in analytical chemistry

Analyses become much more complicated from here, with multiple terms in the equation, more complex hypotheses being tested, and more sophisticated multivariate statistical tools being employed. *However, from the perspective of where zero plays a role, it is the default assumption that there is no relationship between the variables and thus the slope is non-existent or zero.*

Before leaving this section on regression analysis, this is a good point to define a term called *inference space,* which confines the boundary where actual data is used to define a relationship. Anything beyond that is an *extrapolation* that is only valid if the model fitting the data actually holds at higher or lower concentrations. *Interpolation* is when the value of a variable is estimated between existing data points. *The issue of extrapolation becomes extremely important when low concentrations of a chemical are simulated from higher concentration data* (*e.g.*, tissue residues in a food producing animal) *and when cancer risk assessment models and other toxicological endpoints are back extrapolated to concentrations below those experimental data points that defined the relationship.* Figure 4.9 nicely illustrates this issue.

Fig. 4.9 Extrapolation of data beyond the inference space

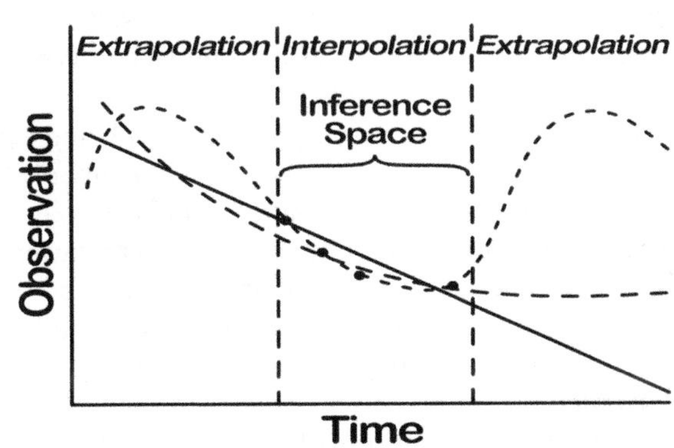

In this example, four experimental data points are available over a time span that defines the actual inference space. This is the limit of the experimental data. Three mathematical models adequately fit these data points within the inference space: a straight line, a slowly declining curve, and a cyclic process. To extrapolate beyond the observed data, the model that mechanistically defines this phenomenon must be known. If additional data is collected at higher or lower response levels, then the correct model would be easily selected. But what happens if data outside of the inference space is not available? This example could be representative of a hormone concentration that has a diurnal (night—day) or monthly cycle (dotted cyclic line) or normal chemical decay in the body governed by different order elimination processes. In this case, only the model marked by the solid line would even approach zero.

Computational Dilemmas Using Zero

We now have a grasp of how the number zero is handled in computational mathematics and statistics. We have defined where its appearance can cause issues in computation. However, once we leave the realm of computation and pure mathematics, we enter the world where these tools are used to address issues in public health and safety, *where zero is meant to reflect the absence of a substance or biological effect.* In this realm, the definition of zero becomes fluid and very dependent upon the context in which it is discussed, as well as the tools used to measure it. As mentioned earlier, in some minds zero has the air of perfection associated with it, with zero contaminants suggesting *purity*. As will be repeatedly demonstrated in the natural world, zero is but an estimate and very dependent upon the factors and definitions used to measure it. It is not an eternal Platonic constant.

In analytical chemistry, instrument sensitivities may range over orders of magnitude. The interpretation of the concentrations measured may similarly vary over orders of magnitude depending upon the nature of the response being studied, the robustness of the model used to make the toxicological estimate, and the user's interpretation of the actual risk involved. Since measuring is involved, uncertainty enters the calculation and as discussed, statistics immediately will change a point estimate of zero to a confidence interval, itself dependent upon the sample size used to determine it and the level of confidence (Type I and II Errors) defined by the individual making the prediction.

In its practical use, zero is thus far from being a point estimate and could appear in the result of a risk calculation based on any of the above conditions being triggered, resulting in the final calculation either going to zero since a product includes zero, a time goes to infinity since zero can never be reached, or the calculation *bombs out* since zero appears in the denominator of an equation. One of the biggest problems encountered in risk assessment applications in pharmacology and toxicology is when a linear regression model is used to solve for an equation, such as the first order rate process discussed above, and the lowest concentrations of a chemical or drug falls below levels of detection resulting in them being coded as 0. This creates a problem for, as we will present in later chapters, many of these first-order rate equations are *asymptotic* with the *y* axis at *0* concentration. They are expressed in logarithms, and the log (0) is indeterminate.

There is one positive property of zero, and that is that zero is a *scale-invariant quantity*. Zero is unchanged when units of concentration or weight are changed; 0 mg and 0 µg are equivalent. When rescaling equations to correct for unit changes, *e.g.*, ppm to ppb or mg to µg, all ordinal values must be adjusted, but 0 remains 0. Mathematically this is true, but practically one gets into trouble because the 0 mg that is determined in Study A may be 100 µg in a more sensitive Study B.

Zero can inadvertently appear for several reasons in any number of automated applications at both the input and output levels:

- A small number is automatically truncated to 0 [0.0009 to 0.000]
- A small number is rounded to zero [0.0002 to 0.000]
- Mistakes in unit conversions (metric to US customary units, intra-metric conversions: mm to m, *etc.*)
- A coded logic statement that if $x \leq 0.001$, then let $x = 0$
- *Python code:* def condition_x(x):
- if x < = 0.001:
- x = 0
- return x
- An analytical chemistry instrument simply cannot detect a signal from its detector and reports out 0 instead of its limit of detection (LOD) or limit of quantitation (LOQ). This will be fully discussed in the chemistry chapter.
- A simple human mistake, either in the chemistry phase (mislabeled tube or mistake in pipetting), data entry phase, coding phase, or data transfer or web crawling process.
- In a single-cell RNA sequencing experiment, a high proportion of genes are zero-expression and thus expressed as zero. This so-called zero-inflation controversy has resulted in techniques being developed for reducing zeros. The issue is in some cases a zero is low to no gene expression, while in others it is missing data. These two scenarios have very different interpretations.

Some of these statements can be found as the output of an analytical chemistry program in its linking interface to an automated clinical chemistry or hospital record systems. They could occur when using a spreadsheet to collate data for input into a computational program. Any program that limits the number of significant digits through truncation can cause this type of issue to occur. They can be found in a machine learning algorithm that is *scraping* the web for concentration data in published literature or government reports. One should note that each of these reports posted on-line could already have implemented their own selection cut-off algorithms, rules that are often difficult to find even when one is looking for them! They are often not directly linked to the data that is scraped from the web, meaning there is not a meta tag stating the definition used. Transparency is crucial as more data is released into the cloud.

To prevent this from occurring, some software will automatically report out the LOD or LOQ as the actual observed concentration, without any indication if this *was* the LOD or LOQ and *not* an actual concentration within range of the calibration curve. This is often seen when analytical systems are directly integrated into laboratory management software or public health monitoring systems. This is designed to avoid zero ruining a computation, but it adds a value that has much less certainty than an actual concentration measured within the range of a well validated calibration curve. LOD is less than LOQ and just this simple choice could dramatically alter time estimates of a concentration to fall below these levels. Furthermore, as will be seen in the chemistry chapter, LOD and LOQ are *experimental descriptors*, being a function of the number and range of control samples used in the calibration curve and the specific assay and instrument being used to measure the chemical. *It is NOT a value independent of the assay, the detector, and its calibration curve.* Yet I have seen LOD and LOQ in tables assigned to a chemical concentration threshold without any mention of the assay being used. This is wrong. When the analytical method improves, LOD and LOQ get smaller, yet the concentration is still non-zero.

Such selections of a low but non-zero number may also be made by a scientist who adds an assessment of risk to the concentration obtained. For example, the measured concentration may be an order of magnitude less than what is defined by a regulation as say the "lowest level of regulatory concern", thus it is reported at this threshold and then defined as zero from the perspective of that risk assessment. However, the concentration is actually non-zero, and if new criteria for the risk threshold are determined, the value may be non-zero, but is often lost because the raw data is not available. These issues will be fully discussed in later chapters.

Summary and Moving Forward

We now leave the worlds of history, philosophy, and mathematics to venture forward to examine what zero means applied to the real world. This discussion will cover studying what is the smallest thing possible in our universe and what is the lowest temperature achievable. These and similar concepts fall into the realm of physics and cosmology. In chemistry, we will carefully examine how the presence of a substance is detected and quantitated. From that point on we will deal with applications applied to areas of biology, medical science, and toxicology. We will take a deep dive into the practice of risk assessment in the twenty-first century. All these scientific disciplines employ mathematical models and make statistical inferences from real data, and often generate zero as a result, thereby using all the concepts and tools just introduced in the present chapter. Where necessary, we will present the actual equations and statistical models employed. The author will rely on the reader to consult back to this chapter for definitions and computational issues arising from dealing with zero.

Selected Readings

Darling D. *The Universal Book of Mathematics: From Abracadabra to Zeno's Paradoxes.* New York, Wiley, 2004.

Environmental Protection Agency. *Risk Assessment Forum Whitepaper: Probabilistic Risk Assessment Methods and Case Studies*. Washington, *DC,* EPA/100/R-09/001A.

Haastie T, Tibshirani R, and Friedman J. *The Elements of Statistical Learning: Data Mining, Inference and Prediction. 2nd Ed.* New York, Springer, 2009.

Jiang R, Sun T, Song D, and Li J. Statistics or Biology: The Zero-Inflation Controversy about scRNA-seq data. *Genome Biology* 23, 31, 2022.

Kaplan R. *The Nothing That Is – A Natural History of Zero.* Oxford,Oxford University Press, 1999.

Rawlings JO, Pantula S, and Dickey DA. *Applied Regression Analysis: A Research Tool.* 2nd *Ed.*, New York, Springer, 1998.

Spivak M. *Calculus on Manifolds: A Modern Approach to Classical Theorems of Advanced Calculus.* New York, Addison-Wesley, 1965.

Stewart J, Watson S, and Clegg DK. *Multivariate Calculus, Metric, 9th Ed.,* Boston, Brooks/Cole Cengage Learning, 2020.

Wackerly DJ, Mendenhall D, and Scheaffer RL. *Mathematical Statistics with Applications, 7th Ed.,* Boston, Brooks/Cole Cengage Learning, 2011.

Chapter 5
Physics

In science there is only physics; all the rest is stamp collecting. (Lord Kelvin)

Now that we have a reasonable idea of the history and meaning of nothing and how zero is entered into mathematical equations, this chapter will explore the physics behind defining zero and coming to grasp with the physical conception of nothing or emptiness. This issue can be a moving target as technology has improved our ability to measure increasingly small and faster phenomena. In fact, as I was writing this chapter, the 2023 Nobel Prize in Physics was awarded to three scientists *(Pierre Agostini, Ferenc Krausz, and Anne L'Huillier*) who developed methods to generate light pulses at *attosecond* (one quintillionth or 10^{-18} of a second) durations. To understand how short this really is, an attosecond is to a second as a second is to over 30 billion years! However, the real limit as to what zero means in physics is not instrumental detection, but instead limits based on theoretical extrapolations based on existing models. There are many physical phenomena where the void or zero play a role; *is the void or truly empty space possible, what is the lowest temperature that can exist, is there a zero time?*

As stated in previous chapters, the author will limit discussions to the minimum necessary to define these issues and trace recent advances. Physics plays a central role in what is the practical scientific basis and appropriate contexts for defining the smallest amount of some form of matter existing, with whatever existing below this value being defined as zero or nothing. In later chapters where zero is used in more applied sciences and regulations, our definition will generally be *operationally defined* in the context of either analytically undetectable concentrations of a chemical or by determining what concentrations are low enough to not cause a biological or toxicological effect in a specific system. This often occurs when tolerances based on safety are established for chemical residues of certain compounds. Additionally, recall from the last chapter that assigning a value of zero to any quantity that inadvertently, carelessly, or mistakenly throws zeros into equations out of their contextual domain, can result in significant issues.

Figure Use Attributions
Figs. 5.1, 5.3, 5.5, 5.7, 5.8, 5.9, 5.10, 5.11, and 5.12: Wikimedia – No restrictions
All other figures are either self-drawn, photographed or from author's own publications (attributed)

J. E. Riviere, *Zero – Much to Do About Nothing?*,
https://doi.org/10.1007/978-3-031-82998-7_5

Physics provides us with an insight into addressing both the topics of nothing and zero since one broad goal of physics is to define the structure of the universe. More importantly, physics provides an elegant definition of the different *scales of matter and space* and *why laws and theories applicable to one scale may not be applicable to another*. To use the tools and theories of physics, we have to define a perspective or frame of reference as to how we will interact with nature relative to the definition and properties of particles and matter.

In modern physics today, concepts such as quantum fields, are used to describe the underlying fabric of space. *Fields, rather than particles, are the fundamental building blocks of matter.* One must stress that although we will call our fundamental building blocks of matter *elementary particles*, they are in fact not really particles but structureless points where concentrations of mass, charge, and spin reside. Fields continuously extend through space, a theory consistent with some of *Einstein's* original thoughts in his *General Theory of Relativity* where space could be considered as a type of universal substrate that can bend and fold. Fields are necessary to achieve what physicists term locality which defines the regions of subatomic space where we experience particles. Quantum fields produce particles with some particles like photons having no mass (Fig. 5.1).

Fig. 5.1 Albert Einstein

A basic tenet of quantum theory is that particles can also exist as waves (Fig. 5.2). This particle-wave duality reflects the concept of *complementarity* which suggests that a single phenomenon when considered from different perspectives, or measured using different techniques, may seem to have different or even contradictory properties. *The question one asks defines the concepts and methods one uses. For the purposes of this book, this fundamental property of matter has little impact discussing particles or defining empty or zero. Subsequent discussions we will deal with are processes at a scale where particles are the important components.*

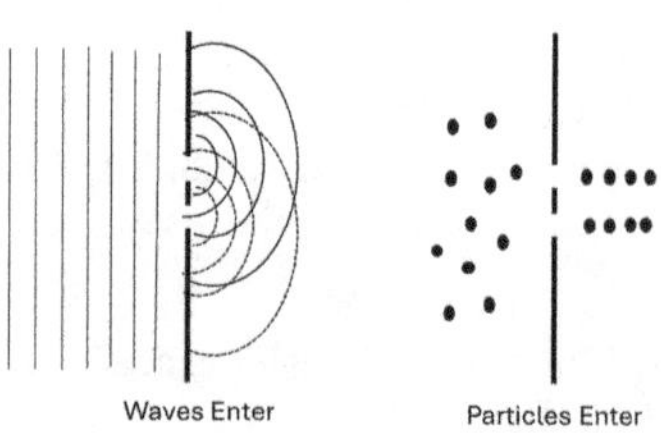

Fig. 5.2 Particle—wave duality

In physics, zero can represent the absence of any physical quantity—force, substance, or temperature. Compared to other more applied sciences where zero is often defined by instrumental barriers to detection and quantification, physics allows one to extrapolate downward in scales under ideal conditions, to explore what zero or nothing would really mean in the physical universe as we now understand its laws and structure to be.

We will discuss what modern physics has to say about whether truly empty space devoid of particles exists in the universe, and then try to define the smallest scale of objects that make up matter; that is the *scale of the granularity of matter,* as well as the lowest temperature that can occur, and the shortest time interval possible. *These values would be considered the smallest unit possible in their respective domains, the physical equivalent of one, below which would be defined as zero or nothing. Zero would be that amount indistinguishable from nothing*. Finally, we will discuss what many consider the big question, was there a time zero denoting the birth of the universe and did the universe always have physical extension, or did it start as a singularity of zero dimension?

Space, Scales and Matter

Is there empty space in the universe? What provides us with the perception that space has expanse and dimensions, and that space is what we move through? We discussed the thoughts of ancient scholars that ranged from the impossibility of the void to concluding there is empty space between the smallest particles of matter. Modern quantum physics and cosmology may shed light on this question through phenomenon such as quantum fluctuations and quantum voids. We then will attempt to define the "real matter or substance" studied in the remainder of this book and estimate the lower limit of its dimensions.

Of all of the sciences, physics provides us with the best perspective of how empty our universe really is. *Starting from infinitely large to subatomic scales, most of the space in our universe is empty, devoid of particulate matter (e.g., atoms and molecules) as we understand and define it.* For the purposes of this discussion of empty space, again we will minimize discussion of concepts such as quantum energy fields, dark matter, and wave-particle duality and instead deal with matter defined and observed as particles. At the quantum scale, fields and forces rule, but *at the atomic, microscopic, and macroscopic scales discussed below, particles are assumed to be the manifestation of matter that causes physical effects and are the objects which manifest action at these higher scales.* Our goal here is to define empty or the void as 'space without particles', specifically the absence of atoms and molecules.

So, is there empty space in our universe? There is empty space occupying thousands of light years between galaxies, sparsely filled with the occasional hydrogen atom or roving star. As a note, the speed of light, denoted as c (299,792,458 meters/second), is used as a standard in physics since c is constant (at one's locality) and is the upper speed limit for movement in the universe. The distance that light travels in one year is the unit of the cosmic ruler termed a light year (approximately 9.46 trillion kilometers).

This empty or so-called free space in deep interstellar space has a particle concentration of roughly a few hydrogen atoms per cubic meter. The total emptiness of space was recently confirmed by the New Horizon spacecraft flying some 5.4 billion miles from Earth. Scientists for a few hundred years have tried to create vacuums in laboratories by pumping air out of glass enclosed containers. Figure 5.3 shows an early version of such an apparatus. The vacuums have improved along with advances in pumps, membranes, and tank materials. The best laboratory vacuum chambers can achieve about 100 particles per cubic centimeter, which is one million times less than that seen in deep space. *Nothing* in a laboratory vacuum is very different on earth compared to deep space.

Fig. 5.3 Early laboratory vacuum chamber where air was evacuate by a hand-pump

Within galaxies, stars are many light years apart. This is clearly appreciated today when viewing the magnificent photos that the Webb space telescope generates. When one enters a solar system, most of the volume is empty except for the star (*e.g.*, sun) and its emissions (such as our solar wind), along with the star's widely separated family of planets, planetoids, asteroids, comets, and some relatively higher concentrations (compared to intergalactic and interstellar space) of molecules and other stuff that is drawn to or trapped there because of their attraction by the gravitational field of the local star.

If we now switch from telescopes to microscopes, we see what we experience as hard matter that has a range of densities and weight that humans can perceive (and sit on if we are looking at what makes a wooden chair hard). But when examining this material at higher magnification, it is made up of atoms that are composed of relatively dense nuclei surrounded by electron clouds in orbitals far removed from their nuclei. Most of this atom which we consider solid, is again empty space (approximately 99.9999999999996% of its total volume). Put into a human perspective, a proton in a hydrogen atom is like a dime in the center of a football stadium.

As discussed in earlier chapters, Democritus and most early scholars considered the conceptual atom (they had no idea of its subatomic properties) as the indivisible particles of which everything else is composed. We now know the atom itself is composed of what we refer to as *elementary particles*. What is the smallest particle detected to date? A first look shows an atom being composed of relatively large protons and neutrons (nucleons) that comprise the nucleus, surrounded by tiny orbiting electrons (when detected and described as particles) having 2000 times less mass than the nucleons.

If we slide farther down the scale, we discover the protons and neutrons are themselves not fundamental particles but are composed of exotic matter such as quarks. Quarks, if they actually have a size, are about one millionth the size of a proton with six primary quarks (*u, d, c, t, s, and b*) themselves differing in mass by almost 2000-fold. The next smaller scale down enters the realm where particles may be one dimensional loops of energy vibrating in eleven multidimensional strings defined in various forms of *string theory*. Depending on the theory describing such ultra-minute particles, some other exotic entities, such as gravitons, occur.

If we combine these microscopic and cosmic scales of objects in the universe that scientists have actually studied, we have a universe ranging in scale from 10^{-17}—10^{-18} meters to 10^{26}–10^{27} meters, *a scale difference of* 10^{43}–10^{45}. This is 10 followed by approximately 40 zeros! If one steps back and lets the magnitude of these scales "sink in", we realize that most of what makes up the space of the universe from the subatomic to galactic scale is devoid of particles and is truly empty and void.

Thirteen *scales of nature* have been defined by various workers:

1. *Quantum—Fundamental elementary particles*—(Planck lengths)
2. *Nuclear* (femtometers)
3. *Atomic* (angstroms)
4. *Molecular* (nanometers)
5. *Microscopic* (micrometers)
6. *Macroscopic* (millimeter to kilometers)
7. *Sub-planetary* (few hundred kilometers)
8. *Planetary* (1000–200,000 kilometers)
9. *Stellar* (0.08 to 2000 times size of sun)
10. *Stellar systems* (2 light years) 1 light year = 9.46 trillion kilometers,
11. *Galactic* (100–1000,000 light years)
12. *Cluster and void scale* (10–100 million light years)
13. *Cosmic—Observable universe* (93 billion light years)

The number of thresholds and dimensions of such classification scales are very subjective and dependent upon the perspectives of the workers using them. Many scientists collapse these to subatomic, atomic, molecular, microscopic, macroscopic, and cosmic; with boundary size cutoffs (⇑) seen in Fig. 5.4 below often being different.

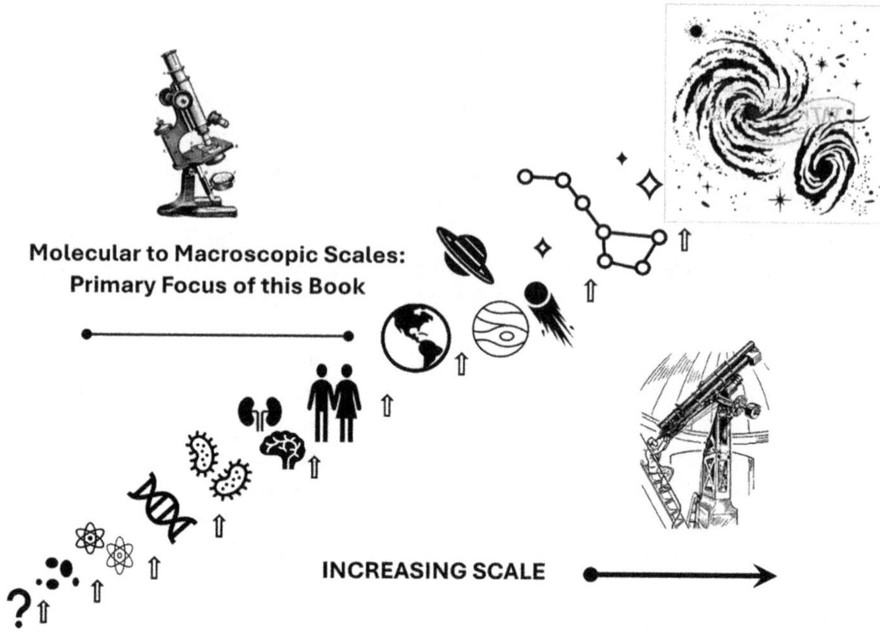

Fig. 5.4 The scales of nature

The importance of such scales to the focus of this book is that *zero must be defined relative to the scale of nature that is relevant to the question at hand.* Each scale is described by its own physical laws, ranging from quantum mechanics to molecular chemistry to biology, genetics and physiology to geology and meteorology and then astronomy and cosmology. At the quantum scale, fields and forces rule but *at the atomic, microscopic and macroscopic scales, particles are assumed to be the manifestation of matter that causes physical effects and are the objects which manifest action at these higher scales.* Our *goal in this book is thus to define zero where there are insufficient particles to be detected or cause an action to occur in the relevant domain. These concerns range from molecular to macroscopic scales.*

As we move across scales, laws that are needed to describe or predict behavior in one scale will not be relevant in the other. This concept of *scalar dimensions* is particularly important since physical phenomena of different length scales cannot affect each other and are said to decouple. However, *emergent behavior* occurs where systems at larger scales behave according to their own properties that are not present at the lower scale but emerge from the properties of the matter and energy below them. Good examples of this are seen in large networks composed of relatively simple components that function at a much higher level of complexity than an individual component could accomplish. Life is an emergent phenomenon with cells and organisms possessing properties that are far more complex than their individual components. Another excellent example of emergent behavior is the *Periodic Table of Elements*. Somehow the complex properties of all these elements which we will discuss in the next chapter emerge from three ingredients at the atomic scale: electrons, nuclei of different charges, and the Schrödinger or Dirac equations. A classic biological example is the complex social behavior arising from individual ants or bees residing in a colony and our brain's complex behavior arising from a network of billions of individual neurons and trillions of their synaptic connections. Some consider the universe itself to be an emergent phenomenon.

These different laws are related to the objects that comprise that scale and forces that dominate the structure of the objects in that scale's domain and the interactions between them. In general, the strength of all interacting forces is inversely proportional to the square of the distance between the objects. For the subatomic and atomic scales, strong nuclear forces that operate over very short distances dominate. As we get to molecular interactions, forces that form covalent bonds between atoms and those such as ionic attraction, hydrophobicity, hydrophilicity and *van der Waals* forces modulate interactions between molecules at nanometer distances. These forces themselves are not independent as, for example, hydrophobic forces emerge from van der Waals and electrostatic forces combined with entropy. However, it is these higher-level inter-molecular forces that define the domains of biochemistry, as well as molecular pharmacology and toxicology.

As will be extensively discussed later, *zero applied to this domain is most relevant for defining minimum detectable concentrations as well as the minimum mass of substances that cause perturbations to chemical and biological systems.* At macroscopic scales, physical forces such as fluid and air flow, pressure, diffusion, osmosis, and macroscopic structures (membranes, intracellular compartments, cellular cytoskeletons, transport systems, enzymatic pathways, receptors, types of cells, blood vessels, etc.) dominate many biological interactions. Some view this as the *mesoscopic* scale of few hundred nanometers at the interface between the microscopic and macroscopic domains being the most important realm for studying biochemical and biological interactions and provides the basis for nanoscience. Some forces can operate over many scales, with magnetism being an excellent example. At the next larger scale, ecologists define scale in terms of grain and extent.

Interesting things often happen at the mesoscopic scale where forces from both the smaller and larger scales may have to be taken into consideration. Physicists use the technique of renormalization to relate the multiple scales of nature by redefining variables in equations to eliminate inconsistencies that have no relevance to the problem being studied at the next larger scale. Their concept of an *interpretive ladder* is used to climb from simpler objects to more complex ones. A good example of the need to independently deal with properties across scales that the author has encountered in his research is the weird science of nanotechnology. Quantum mechanics is needed to describe the properties of the materials comprising a nanomaterial (*e.g.*, quantum dots, carbon nanotubes, *etc.*) but molecular interactions define how the nanomaterial will interact in the biological domain.

Now that we have discussed what matter might exist in space at different scales, and have discussed what the smallest particles may be, what would we consider zero distance? In our macroscopic biological world, the unaided human eye has an angular resolution of about 0.02 degrees and can discern objects down to 40–50 microns. As can be appreciated from the earlier discussions, the true smallest distance in the physical world would be related to the granularity of space, an area that quantum pioneer Professor *Max Planck* worked on at the turn of the twentieth century. This dimension has been defined as the *Planck length*, 1.6×10^{-35} meters, though there is some debate that it could be 3 ½ times larger depending on the theory used to estimate it. It is a limit on the precision with which we can probe and describe the atomic and microscopic worlds as we simply do not understand what happens at and below this scale. The Planck length is trillions of times smaller than any instrumental approach we have to measure it. It also is the diameter of a black hole singularity, roughly 10^{-20} times the size of a proton! (Fig. 5.5)

Fig. 5.5 Max Planck

This *granularity* of space is the smallest measurement of length which possesses any meaning, below which our ideas of space-time and gravity cease to be valid (since we do not yet have a quantum theory of gravity that is needed to move downward). If our universe is truly distinct at this scale, then this would be *the minimum physical distance* at which we can make reasonably reliable statements, below which would reside *zero physical distance.* Discussion at this scale must be considered speculative. If we recalculate the extent and scale of the universe, as we did above based solely on particles, but instead start at Planck's length, our existing science can describe a universe extending from 10^{-35} to 10^{27} meters, a scale difference of 10^{62}.

For our purposes, this defines the smallest scale possible when we talk about the granularity of space itself; a number still orders of magnitude below the size of quarks. One can also conclude based on modern physics, when viewed as a whole, our universe is pretty much empty and void of matter. Empty would be defined as space devoid of the smallest particles we can detect. However, is this empty space truly empty? As mentioned above, research in quantum mechanics suggests otherwise and hold that all of space-time, including our now defined empty space, is perfused by force fields (gravity, Higgs field, etc.) and electromagnetic energy waves including high energy gamma radiation, from where quantum fluctuations in space-time can occur allowing matter to appear from nothing. In field theory, space itself has a zero-point energy (see temperature discussion below) thus truly empty space cannot exist as there is always energy. Field theories are continuous functions where at some points in their probability density function, local bundles or eddies occur and condensed to become what we define and model as particles.

These topics in quantum physics are far beyond the focus of this book and many details have not been agreed upon as quantum theory and general relativity have yet integrated. This author assumes that based on our knowledge and state of science today, we have for our purposes reasonably reliable physical definitions of the smallest particles, particle-devoid empty space, as well as what a distance of zero would mean if our universe is truly granular.

Temperature and Thermodynamics

The next topic in physics deals with heat and temperature, a topic introduced when we discussed thermometers in the introductory chapter. Temperature is basically a measure of particle motion which generates thermal energy or heat. Heat is the form of energy transferred between two substances of different temperatures. Heat transfers from regions of high temperature to regions of low temperature. At high temperatures, particles move and create pressures (manifested in the work of steam engine) and slow down at lower temperatures (think of your hot mug of coffee cooling off on a cold winter day). Chemical reactions speed up at higher temperatures. These are the processes which also occur at the scale of living organisms.

Absolute zero in the Kelvin scale is the lowest limit of the thermodynamic temperature scale, its origin point. Unlike Fahrenheit and Celsius scales, there are no negative temperatures below it. Absolute zero is rigidly defined as *the state at which the enthalpy and entropy of a cooled ideal gas reach their minimum value, recorded as zero degrees K.* Based on international agreement and determined by extrapolating the ideal gas law, *absolute zero (0^0 Kelvin) is – 273^0 Celsius or – 459.67^0 Fahrenheit.* At this temperature, basic particles have minimum vibrational motion, they are still. They retain only quantum mechanical zero-point energy-induced particle motion. In quantum mechanics, matter at absolute zero is in its ground state, the point of lowest internal energy. At temperatures approaching absolute zero, matter exhibits quantum effects such as Bose-Einstein condensation, superconductivity and superfluidity. Pure substances can form perfect crystals with no structural imperfections. Quantum mechanics says that no matter how you try, you can never remove all of the motion from the system. From a quantum pendulum, there is motion and some kinetic energy in the ground state. In this strange world at absolute zero, there are also possibilities of negative temperatures and in fact if you approach zero from the negative side, it would be the hottest temperature. But let's return to our normal balmy world at temperatures much higher than absolute zero.

This is a good juncture to discuss the science of heat and temperature, the science of *thermodynamics,* which governs most of our chemical and biological processes at the scale where life functions. The *Laws of Thermodynamics*, first based upon the work of *Carnot* in 1824 and modified by *Clausius* and *Thomson* in 1860 and finally by *Nernst* in the early 1900's, are numbered starting with the first law, interestingly for this book, being zero.

- The *Zeroth Law* defines thermal equilibrium and provides the basis for the use of temperature as a parameter of thermodynamic systems.
- The *First Law* is conservation of energy within a closed system defining work as the means to transfer energy in and out of the system.
- The *Second Law* describes the irreversibility of natural processes and the tendency of processes to lead to homogeneity of matter and energy and decrease in structure and order.
- The *Third Law* states that a system's *entropy* approaches a constant value as its temperature approaches absolute zero. This absolute zero can be defined as a system in its ground state, with minimal thermal energy having only one configuration.

From the perspective of our discussions, the laws of thermodynamics are important as *entropy defines the arrow of time as moving from states of low (ordered) to high states of entropy (randomness and disorder)* (Fig. 5.6).

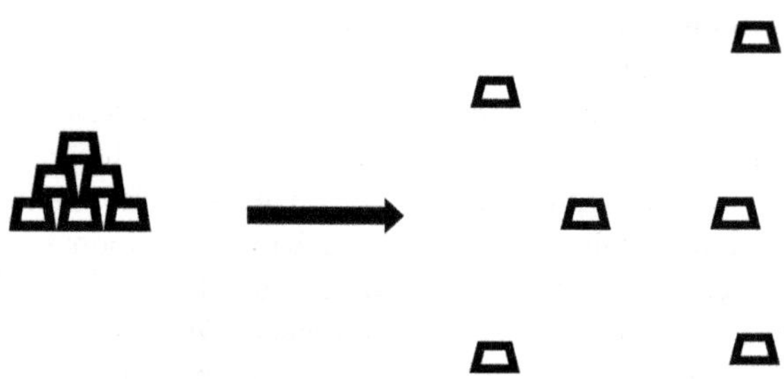

Fig. 5.6 Movement from order to disorder illustrating entropy and time's arrow

Thermodynamics assigns a value to the entropy at a singularity as having the highest amount of entropy for a volume its size (one Planck Length wide). The black hole singularities can accrete and engulf other matter consistent with the second law because entropy would have increased. Some consider black holes as spheres of maximum entropy.

Time, Clocks, Nerves, and Vision

Now that we have defined the physical scale of zero distance and absolute zero, our next challenge is to tackle zero time. But first, what is time? There are numerous perspectives upon which to base this definition. *Consistent with the focus of this book, we will look at time that is relevant to our experiences in the microscopic and macroscopic domains that biochemistry and biology reside in.*

From the perspective of fundamental physics and the theory of relativity, time is not distinct from the other three dimensions of space (length, width, and height) and can be considered part of a four-dimensional space-time continuum. What we consider our three-dimensional space is actually an illusion. *Einstein* expressed this succinctly as *Time and space are modes in which we think and not conditions in which we live.* We will not follow this path further as it does not directly impact on the focus of our discussion. Physicist *Lee Smolin* wrote that *space is nothing but a network of evolving relationships. These relationships are not among things situated in space — they are among the events that make up the history of the world. The relationships define the space — not the other way around.* In humans, the current moment will become etched in memory and defined as past, although cognitive neuroscience has shown that these stored memories can become altered as our consciousness tries to place past events in its ever-evolving self-written narrative, the story of one's life.

Time defined by thermodynamic entropy largely creates the *arrow of time* where the past was more orderly (low entropy), and the future is more disordered (higher entropy). As the universe expands, there is greater space for states of matter to expand into. As entropy expands, we experience the arrow of time.

Classical Newtonian physics describes motions of objects on our human scale to high accuracy. In our everyday chemistry and biological experiments occurring at a scale that molecules and cells are constructed, and at temperatures compatible with life; time and motion are inseparable. Its descriptive adjectives use terms such as velocity and acceleration. As discussed in the overview of calculus, derivatives are instantaneous rate equations (dx/dt) where dt approaches zero. We will attempt to define what instantaneous means using modern instruments.

Biological time can also be linked to the rising and setting of the sun. Its passage during the day was recorded using the simple sun dial (Fig. 5.7).

Fig. 5.7 Sun dial recording time as movement of the sun.

Mechanical clocks conceptualized time in human minds as having the distinct intervals of the ticks of a clock or the hands of a clock similarly jumping every minute. There is no doubt that this perspective shaped human perceptions of time as discrete *packets* rather than a continuous flow (Fig. 5.8).

Fig. 5.8 Mechanical clock ticking away time

Finally, psychological time in the human mind requires the flow of time to be linked to memories that compare the present to the past. In many respects, the perception of the passage of time to humans is nonlinear and varies as a function of overall activity, exposure to drugs and often with increasing age. Psychologists and brain researchers believe that our perception of the present is really the cognitive experience of now. The past is sensed through stored memories and our sense of the future is based on the brain's predictions of what will happen in the coming seconds or hours. The flow of time is thus an illusion with time being driven by our internal biological clocks, which are susceptible to numerous sensory inputs and states of consciousness driven by input from numerous brain centers.

How precise are modern instruments in measuring time? Recall that at the beginning of this book, the 2023 Nobel Prize in Physics was awarded to scientists who developed methods to generate light pulses at attosecond (one quintillionth or 10^{-18} of a second) durations. Currently the shortest duration of time ever measured was a zeptosecond (10^{-21} s), the time it took for an electron to escape an atom during photoionization. However, at the macroscopic- and biological- scaled world we inhabit, *30 milliseconds* is the shortest time interval our unaided human sensory system can discern change at the receptor level. For example, this would be when a photosensor in a human retina detects a flash of light. These sensory signals must then be transmitted by neurons to the brain and then be processed there to consciously register a change. Neurophysiological studies have shown that a *few hundred milliseconds* are the duration of the fastest time our brain can discern that a change has occurred in our environment after cognitively processing the sensory input. At this scale, this would be defined as *instantaneous* to the unaided human observer (Fig. 5.9).

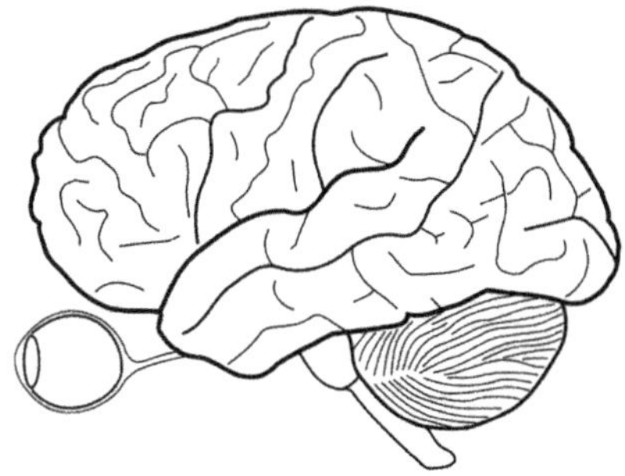

Fig. 5.9 The eye and the human brain

The processing of sensations by the nervous system can sometime avoid the delay inherent in notifying the conscious brain that something has happened. In cases of touch, pain, and temperature sensation, in addition to sending signals to the central cognitive nervous system, there are also shorter direct feedback loops through the spinal column that will activate motor neurons to remove the affected body part from the object that caused the sensation (a nail, a hot iron, *etc.*). In neurophysiology, there is a physiological recovery period where the neuron must also reset its ion channels (the basis of nerve signal propagation) after a signal has passed and where chemical synapses connecting neurons must reset. When you take all these considerations into account, one can appreciate why our brain can only discern differences in repeated visual images of about 40 frames per second. This *visual frame rate* is the limit that we can process visual signals, a phenomenon that enables us to watch movies that are just in reality a parade of still images. Our other senses; taste, hearing, smell, and touch have similar neural transmission and cognitive delays that define their own frame rates.

Returning to physics and following the same approach we did for defining zero distance, we could define a physical theoretical shortest time interval as the time required for light to travel through empty space a distance of one Planck length. This is approximately 5.39×10^{-44} *s*. This value is termed the *Planck time* (t_p)*, an interval 23* orders of magnitude shorter than the zeptosecond duration currently detectable by our modern instrumentation. Any time shorter than this would have to be defined as zero time.

It must be stressed that when we discuss theoretical zero distance or zero time, we are many orders of magnitude below what workers are able to even measure using current technology and instrumentation, forgetting that our brains could never keep up with such short intervals. As will become a recurring theme of this book, *zero should be defined at both the physical scale and context of the question being asked at hand.* It is simply incorrect to use definitions from a lower scale of nature to interpret data generated from a higher scale.

Miscellaneous Zeros Including Resolution of Vision, Sound and Hearing

While we were discussing limits of vision relative to speed of neuronal transmission, we have not defined limits of visual resolution of the human eye. What is the scale of pixels in human vision? The theoretical maximum angular resolution for the human eye, based on the Rayleigh diffraction limit for light entering the pupil, is about 28 arc seconds, which equals an angular resolution of 0.008 degrees. This is about 140 mm seen from a distance of 1 km. Of course, this is the best human vision, most of us operate at about 1 arc minute (60 arc seconds) or 0.016 degrees. The lowest intensity of light that has been reported detected is about 50 to 60 photons. The absolute threshold of vision has been reported as typically around 0.01–0.001 candela/m^2. Anything smaller than this resolution or of less intensity would not be perceived, such a signal or object would be nothing!

Animals and humans also have very different abilities to see light of different wavelengths. The full electromagnetic spectrum is shown in Fig. 5.10. Human color vision is a result of possessing three types of visual receptors or cones (red, blue and green) allowing us to see colors from red through violet (≅ 400 to 700 nm, 4 to 7×10^{-7} M). The names of the wavelengths above and below our visual range (ultraviolet and infrared) reflect the bias given to the range a human can see. Some animal species including dogs and cats, have only two types of cones (blue and yellow) making them similar to color-blind humans. Some birds and fish may have up to 4 or 5 different color receptors, making their perception of the world much richer than ours. Some animals see below our visual spectrum into the infrared while others can perceive ultraviolet light. Because heated bodies radiate infrared light, many predators are sensitive to this spectrum that facilitates night-vision and some, such as pit viper snakes, have evolved infrared sensors to attack their prey. The interpretation of not seeing a color is very dependent upon the capabilities of the individual and species involved.

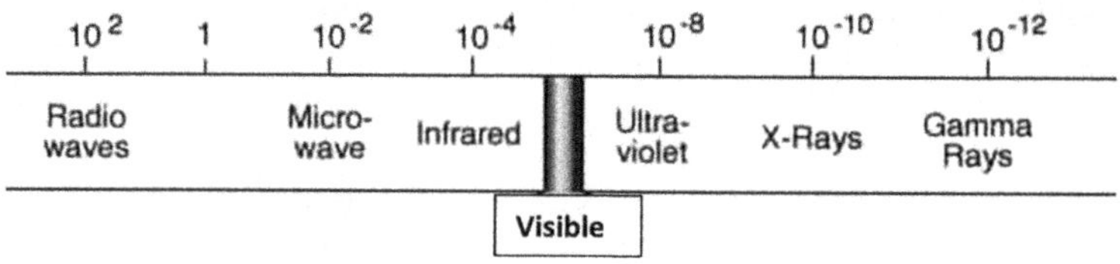

Fig. 5.10 Electromagnetic spectrum (wavelength in meters)

This is a good point to briefly discuss one other area of physics where the meaning of zero has been widely discussed, which is sound. *Sound is defined as a vibration that travels as acoustic pressure waves of high and low pressure through a transmitting medium, such as air, liquid, or solid.* Sound waves are termed longitudinal waves as the vibrating particles propagate parallel to the energy source emitting them. The speed of sound is dependent upon the medium, with *faster* transmission occurring at higher temperatures and *lower* in heavy gases. *In air, at 20 °C it is 343 meters per second.*

There are four properties that describe any wave: frequency, amplitude, wave form, and duration. In sound and music, these are equivalent to pitch, volume, timbre, and duration. Sound waves can be classified into three types based on their frequency measured in Hertz (Hz—number of cycles per second) that humans can hear: *audible* are those that can be heard, *infrasonic* are too low-frequency (below 20 Hz), and *ultrasonic* are too high-frequency (above 20,000 Hz) for humans to hear. Volume or sound intensity is measured in decibels (dB) where the lower limit of *human* detection is defined at 0 dB (Fig. 5.11).

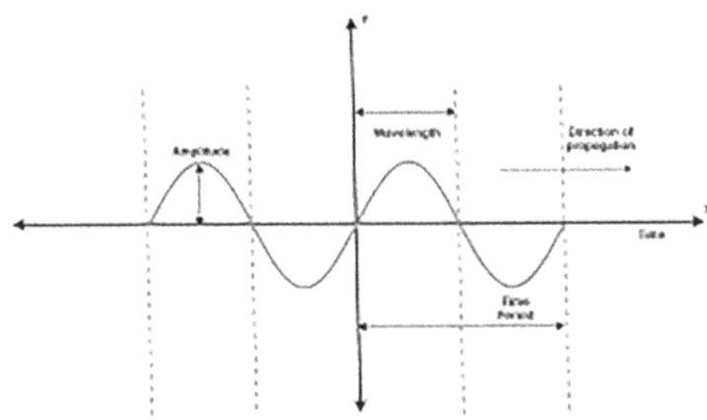

Fig. 5.11 Characteristics of a wave

What does the absence of sound really mean? As can be appreciated from this short introduction, human absence of sound would be the inability to detect sound waves outside the audible range of 20 to 20,000 Hz or which has a volume of ≤0 dB. This sensitivity changes if one is in the water or at high altitudes where air density and temperatures are lower. In a vacuum, a sound wave does not propagate. Thus, the absence of sound to a human is relatively easy to define. However, we know that biological variation will change the ability of some people to detect sound. Other animal species can often hear sounds that either have a volume less than 0 dB or occur at a frequency that we cannot detect. Humans have also developed instruments to greatly increase our sensitivity to detect sound waves. Defining a true zero is very subjective because this would be equal to no movement of air particles or the absence of a medium in which to propagate them. As one tries to detect sounds propagating at various frequencies with very low amplitude, there is an increase of background noise that makes detection of a specific signal difficult. If the frequency is very low, events could be separated at time frames exceeding our observation period.

Sound is a uniquely subjective human experience that we only define because we hear the propagation of pressure waves in the air. Humans have evolved to associate what we call sounds with various environmental signals, such as ocean waves crashing or leaves rustling in the wind; or alternatively as means of communication such as warning cries, drumbeats, or speech. The physics describing sound wave propagation is no different than that of electromagnetic, ocean or even gravitational wave propagation. The difference is that we define sound as those air waves that are perceived by the human ear, given them function and meaning, for example perfect silence in the absence of any detectable noise! In a similar vein, light outside our visual range of detection equates to perfect darkness. Sound waves and photons may still be present but below our biological sensor's ability to detect them, the absence of signal then being cognitively associated with silence or darkness.

Cosmology

The external world as we understand it is an elaborate intellectual construction. Our terms for zero and nothing are embedded in this construction. When the underlying model changes, so do our definitions that were dependent on the original structure. In constructing such models, we practice radical conservatism that requires us to learn from nature and accept certain facts, which are all components of the scientific method. Our direct experience gives us the concept of an extended space as a bin where nature can deposit objects.

Is there a time zero when our universe began? *Saint Augustine* in *Confessions* written around 400 CE, *believed God* created the universe. Before creation, there was no time as time is a function of the universe. St. Augustine considered God as timeless and dwells in eternity where there is no sense of past, present, or future. As we discussed in the earlier chapters, Buddhism has relatively modern concepts that easily deal with complementarity as well as the singularity, as clearly elucidated in the *Dali Lama's* book *The Universe in a Single Atom.* He is comfortable with the disparity between our common-sense perspective as to what the world we perceive *versus* the philosophical and spiritual concepts such as emptiness. The concept of the singularity at the beginning of time is consistent with some modern views of it being a thermodynamic instability as well as the cyclical nature of cosmic rebirth and death (Fig. 5.12).

Fig. 5.12 St Augustine and the Dali Lama

The modern concept of a big bang that started our universe is not an explosion emanating from a single point of infinite mass, but rather the expansion of the universe itself. Modern cosmology gives us the best estimate of time zero as occurring some 13.8 billion years ago, with the universe expanding a billionth of a billionth of a second from being a singularity to a universe where space has expanse. This time is our already encountered *Planck Time*. Laws that describe time going forward from this point reflect the universe moving up time scales, with each "birthday" resulting in a new set of laws to describe that scale's behavior. Physicists have written many volumes on how the universe evolved in these very early time periods, for example, the phenomenon of inflation occurring just after the birth of the universe. Again, these are beyond the scope of this book since we have met our mission, define time zero. As discussed earlier, our understanding of physics before Planck time is also nothing! What could have happened before the singularity pretty much follows the thoughts of the Dali Lama. If there was always only one universe, then time began when the singularity expanded. In this context, time could be considered an emergent phenomenon embedded in quantum entanglement within our space-time continuum and does not exist outside of it. In other models, the universe expansion and contraction are cyclical with our singularity being the crush point resulting from the universe before us.

There are some variations of this theory including *Julian Barbour's Janus Point*. I mention this solely because in its formulation, time zero at the Janus Point becomes so important that he refers to it as the *Royal Zero*. This, for our favorite number zero, was coined by one of the giants in quantum physics, *Erwin Schrödinger*, who referred to zero as the number with *Royal Blood*. Barbour sees the Royal Zero as the singularity where the size of the universe vanishes. A fitting topic to conclude this chapter.

Summary

One of the founders of quantum mechanics, *Werner Heisenberg*, stated that *"What we observe is not nature itself, but nature exposed to our method of questioning"*. Physics gives us a framework to define nothing and zero in terms that enable scientific exploration. It teaches us that defining these terms is dependent both upon the methods used to study it as well as the context we intend to use them. The definitions are based on the scale of the phenomena we are studying which define the laws that should govern the system's behavior. They are also dependent on the mental models we construct to explain the observations we make.

We will now move forward and discuss the particles of matter that will be the focus of the remainder of this book, the world of atoms and molecules, and the field of chemistry.

Selected Readings

Barbour J. *The Janus Point: A New Theory of Time.* New York, Basic Books, 2020.
Dalai Lama. *The Universe in a Single Atom—The Convergence of Science and Spirituality.* New York, Morgan Road Books, 2005.
Feynman, RP. *Feynman Lectures on Physics: The New Millenium Edition,* New York, Basic Books, 2011
French AP. *Einstein—A Centenary Volume.* Cambridge, Harvard University Press, 1979.
Griffiths D. *Introduction to Quantum Mechanics*. Cambridge, Cambridge University Press, 2019.
Kaku M. *The God Equation*. New York, Doubleday, 2021.
Mack K. *The End of Everything.* New York, Penguin Books, 2020.
Odenwald S. "The Origins of Time". *Astronomy*, May, 2022
Sain Augustine (Ed. Meconi DV). *The Confessions: Saint Augustine of Hippo.* San Francisco, Ignatius Press, 2012.
Shanker R. *Fundamentals of Physics I.* New Haven, Yale University Press, 2019.
Siegel E. *The 13 Scales That Define Our Physical Universe.* https://bigthink.com/starts-with-a-bang/13-scales-universe/(Accessed 01/21/2024).
Smolen L. *Three Roads to Quantum Gravity*. New York, Basic Books, 2008.
Tyson ND. *Cosmic Queries*. Washington, DC, National Geographic, 2021.
Wilczek F. *Fundamentals—Ten Keys to Reality.* New York, Penguin Press, 2021.

Chapter 6
Chemistry and Limits of Analytical Detection and Quantitation

> Be a physical chemist, an organic chemist, an analytical chemist, if you will; but above all be a chemist. (Ira Remsen)

Chemistry is broadly defined as that *branch of science that studies the structure, composition, and properties of elements, molecules and their derivatives as well as how they transform and interact with one another. Biochemistry* deals with the *chemistry of atoms and molecules involved in living systems.* Continuing with the theme of previous chapters, we will explore what zero and nothing means in this field. As stated often in this book, this chapter is not a textbook of chemistry, and I have limited review of principles to the minimum necessary to define the terms needed for the discussions of zero and nothing. The interested reader should consult suggested readings for further information.

Chemistry is the appropriate domain to start the discussion of what a zero concentration of a chemical means to human health, both from the perspective of the inability to detect that a compound is present and to realizing that there are minimum concentrations needed to interact with biological processes mediated by atoms and molecules. This chapter will primarily deal with the strategies to detect *chemicals,* a chemical being defined as *any substance with a defined composition.* This field is termed *analytical chemistry*. Chemical interactions in biological processes will be discussed in the next chapter on biology and pharmacology.

Chemistry and biochemistry function at the *mesoscopic* scale bridging the atomic and microscopic domains. It is here that the forces governing interaction between atoms and molecules dominate. *Chemists define zero as the absence of atomic or molecular structures, which in analytical chemistry is operationally defined as 'not detectable'. In other words, zero is defined as substances that are either absent or at a concentration below the detection limit.* In chemistry, our discussion stops at the level of atoms, and we will not consider whether other elementary particles or force fields may be present at lower domains. Particles below this detectability threshold may still be present. However, using available instruments, we simply cannot detect them and interactions important to biology and medicine are assumed

Figure Use Attributions
Figs. 6.2, 6.3, 6.4, and 6.5: Wikimedia – No restrictions
All other figures are either self-drawn, photographed or from author's own publications (attributed)

J. E. Riviere, *Zero – Much to Do About Nothing?*,
https://doi.org/10.1007/978-3-031-82998-7_6

to be at the molecular level. In other cases of chemical analysis where the composition of a material is being investigated, zero may just indicate that an element is not present or detected analytically in the sample being studied at amounts similar to other components of the material. With the continuous progress made in scientific instrumentation, the level of detection continues to drop to values well below those with any associated biological meaning. *It is crucially important that the context of how these concentrations will be used and considered when interpreting results from chemical analysis.*

A Short Digression into History and Nothing

In studying the history of chemistry, there are a few points worth discussing relevant to the concepts of nothing and vacuums, as well as the origins of where our present state of knowledge concerning elements and atomic theory arose. Chemistry seems to have its own historic trajectory beginning with the Greek Democritus and his concept of atoms, a beginning in common with philosophy and physics. The term chemistry probably had its origins in Alexandria around 300 CE focused on the art of *alchemy:* the transmutation of metals to gold and the quest of a universal elixir. For the next 1500 years, applied "chemistry" in the guise of metallurgy, dye making, *Materia Medica* (substances used in the composition of medical remedies), and alchemy was practiced. Slowly they yielded the methods, theories, and principles of what would become chemistry today. Alchemy being referred to as the *occult science*, the *pretend science* or the "divine art", allowed chemistry's development to occur on its own. The author contends that this path largely insulated chemistry from the growth of philosophy, astronomy, and natural science touched upon in earlier chapters, along with religious suppression of certain threads of inquiry. For example, in contrast to these other schools, *the concept of particulate atoms was embedded in much of this early work.* Alchemy also flourished in India, China, and the Middle East, with the Islamic philosopher and physician *Avicenna*, whom we met in connection with the void, also making an impact. Throughout this time, the four elements remained largely those of the Greeks: Earth, Air, Water, and Fire. For Hindus, the five elements were Earth (Prithvi), Air (Vayu), Water (Jal), Fire (Agni), and Space (Akasha). Again, Eastern philosophers were comfortable with an *element* that was essentially the void.

In the early sixteenth century, some industrious Renaissance alchemists focused attention on using chemicals, minerals and metals to treat disease. This field was originally termed *iatrochemistry* but today we would refer to it as *medicinal chemistry. Theophratus Bombast von Hohenheim*, known today as *Paracelsus,* is often considered the founder of medicinal chemistry. He laboured in the alchemy paradigm of fundamental elements, his being sulphur, salt, and mercury. His laboratory in Basel, Switzerland is seen in Fig. 6.1. Although the theoretical framework of his remedies was steeped in alchemy and mysticism, he was an astute observer resulting in some of his medicines actually working. We will see Paracelsus again in later chapters since he is also considered the father of Toxicology and the author of that field's nostrum *dosis sola facit venenum* translated from Latin as *dose makes the poison.*

Fig. 6.1 Paracelsus laboratory in Basel (Picture taken by author)

In the seventeenth century more modern perspectives in chemistry started to emerge through the work of chemists such as *Van Helmont* in Brussels. His writings clearly reflected a concept of physical matter where he opined that *nothing is made of nothing, therefore weight is made of anther body of equal weight in which there is a transmutation as it were of this matter*. His work on studying gases *led him to refute Aristotle where he thought a vacuum was far from impossible, but rather quite ordinary*. By the end of the seventeenth century, *Robert Boyle*, whom many consider the father of modern chemistry, firmly planted experiments into the lexicon of the chemist and finally discarded the four Greek elements as being the basis of all matter. He began a path which would lead to our modern understanding of the elements. His work was consistent with the atomic theory as it considered chemical combinations as occurring between elementary particles. These concepts were further developed by chemists, such as *Priestly and Lavoisier*, and ultimately led *John Dalton* in the eighteenth century to formulate the atomic theory.

This rich development of chemistry, rapidly occurring in the intellectually fertile 17th and 18th centuries, firmly planted concepts including atoms and the existence of a vacuum into mainstream science and discarded Aristotelian dogmas that had slowed scientific progress for a millennium. This work led to the modern disciplines of physics, thermodynamics, and the definitions and properties of matter which form the basis of modern chemistry.

Mass and Weight

We will start with the perspective of assessing whether something is present or not based on a particle's mass and weight. Mass is an important physical property of all matter that manifests itself to chemists as weight. We know from the previous chapter, that masses of atoms are dominated by the mass of their nucleus, *i.e.*, the number of nucleons present. The number of protons is referred to as an element's *atomic number* (Z) and defines the element as well as the number of its electrons. The simplest molecule is Hydrogen ($Z = 1$) having a single proton in its nucleus and one electron orbiting it. Helium ($Z = 2$) has two protons and two neutrons in a nucleus orbited by two electrons. Because it has two neutrally charged neutrons, its weight is four times hydrogen. This number, the sum of protons and neutrons, is referred to as the *atomic mass number (A)* and is reflected in the *atomic weight*. An isotope of an element is one that contains the same number of protons, which defines its chemical identity, but has a different number of neutrons thereby increasing or decreasing its atomic weight. As discussed previously as an example of emergent properties, the number of nucleons in an atom will be manifested in the properties expressed by its constituent atoms. This is reflected in the *periodic table of elements* seen in Fig. 6.2. All atoms consist of the same nucleons and electrons, but their number and structure of their electron orbitals and shells result in the emergent properties of the diverse chemical elements that comprise our world.

Group → / ↓ Period	1	2	3	4	5	6	7	8	9	10	11	12	13	14	15	16	17	18
1	1 H																	2 He
2	3 Li	4 Be											5 B	6 C	7 N	8 O	9 F	10 Ne
3	11 Na	12 Mg											13 Al	14 Si	15 P	16 S	17 Cl	18 Ar
4	19 K	20 Ca	21 Sc	22 Ti	23 V	24 Cr	25 Mn	26 Fe	27 Co	28 Ni	29 Cu	30 Zn	31 Ga	32 Ge	33 As	34 Se	35 Br	36 Kr
5	37 Rb	38 Sr	39 Y	40 Zr	41 Nb	42 Mo	43 Tc	44 Ru	45 Rh	46 Pd	47 Ag	48 Cd	49 In	50 Sn	51 Sb	52 Te	53 I	54 Xe
6	55 Cs	56 Ba		72 Hf	73 Ta	74 W	75 Re	76 Os	77 Ir	78 Pt	79 Au	80 Hg	81 Tl	82 Pb	83 Bi	84 Po	85 At	86 Rn
7	87 Fr	88 Ra		104 Rf	105 Db	106 Sg	107 Bh	108 Hs	109 Mt	110 Ds	111 Rg	112 Cn	113 Nh	114 Fl	115 Mc	116 Lv	117 Ts	118 Og
Lanthanides				57 La	58 Ce	59 Pr	60 Nd	61 Pm	62 Sm	63 Eu	64 Gd	65 Tb	66 Dy	67 Ho	68 Er	69 Tm	70 Yb	71 Lu
Actinides				89 Ac	90 Th	91 Pa	92 U	93 Np	94 Pu	95 Am	96 Cm	97 Bk	98 Cf	99 Es	100 Fm	101 Md	102 No	103 Lr

Fig. 6.2 Periodic table of the elements

The periodic table was invented by the Russian chemist *Dmitri Mendeleev* in 1869 as a tool to demonstrate how chemical properties of the elements are a function of their atomic number and can be grouped as blocks, columns, rows, or diagonals in the table. This is a graphical refection of the *periodic law, which states that when elements are arranged in order of their atomic numbers, a recurrence of their properties is evident.* This is because the number of valence electrons in an element's outer shell has a major influence on determining an element's chemical properties because they comprise the interface where interactions occur with other molecules. When there are too many electrons to occupy one shell (2 in 1st, 8 in 2nd, 18 in 3rd), another shell is formed, which starts the repeat of their properties. For example, chemical bonds between atoms involve sharing electrons in the outer shell between atoms. Compounds with different atomic weights may have similar chemical properties if they fall in the same column, examples being helium, neon and argon, or carbon and silicon. The number of orbitals or shells determines the block of rows in which a compound resides. Although fascinating, any further discussion of these properties is well beyond the scope of this book. *The take home for our purposes is that each element has a different intrinsic weight based on the number of its nucleons, and within individual elements resulting in the formation of various isotopes.*

What are the practical implications? If there are two samples of material with the same number of atoms, one made of lead (Z = 82) will weigh more than one made of silicone (Z = 14). Conversely, one gram of lead has less atoms than one gram of salt. Alternatively, one kg of salt has a much greater volume than one kg of lead. When weighing a compound and knowing the reactivity and binding of atoms is related to the number of atoms present, we must have a method to relate weight to number of atoms. This is accomplished by defining a *mole as representing* $6.02214076 \times 10^{23}$ *molecules,* a quantity known as *Avogadro's number.* One mole of water weighs 18 grams while one mole of gold weighs 197 grams. A great analogy would be paper rolls of 50 pennies or dimes; both rolls have the same number of coins but have different sizes and weights, with the paper roll representing the mole and different chemicals being the coins. A role of dimes will have greater impact than a roll of pennies!

Knowing this relationship, one can measure the weight of a compound on a scale and calculate the number of molecules present. Alternatively, if one wants to have equal numbers of different molecules, one can calculate the weight necessary to achieve this goal. This concept is often encountered when solutions are mixed where *molarity (M) is defined as the moles of a compound in a certain volume of solution (e.g. ml or l)* and *molality is the moles of compound per mass of solvent (e.g. g or kg).*

This concept of moles allows a distinction to be made between concentration or mass of a chemical detected in an assay *versus* how many actual molecules are present. This distinction has direct practical application affecting the interpretation of a concentration below the analytical threshold concentration; for the same cutoff for different compounds will reflect vastly different number of molecules being present depending on the size of the molecules being studied. We will encounter moles and molarity again when we discuss chemical and drug interactions in biological systems.

Measurement

Before we discuss weighing and analytical chemistry, it is important to present the experimental contexts involved in making any measurements in a laboratory. Dictionaries and thesauruses list accuracy and precision as synonyms. However, in science, these two terms have very distinct and different meanings. These can be grouped into four types of concerns:

- *Accuracy*: How does the measurement taken reflect the reality of what is actually being measured?
- *Precision*: How reproducible are the measurements being taken?
- *Error*: The difference between measured value and actual value
- *Uncertainty*: The quantification of reproducibility and precision

The statistics discussed in the earlier chapter allows the error and uncertainty to be quantitated using a confidence interval calculated from the samples collected. Accuracy and the resulting error are measured by including known controls into the assay. This could be known standard weights not used in the calibration of a scale or known concentrations of chemical not included in the calibration curve. Precision and its associated uncertainty are measured by using replicate samples and often expressed as a confidence interval. The difference between accuracy and precision is illustrated in the target analogy of Fig. 6.3.

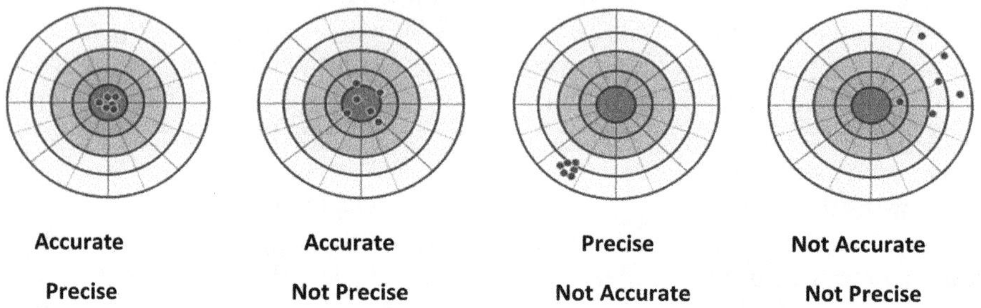

Fig. 6.3 Accuracy *versus* precision

A good example of accuracy *versus* precision is the second target above. Having a thermometer that is off by a few degrees due to mis-calibration showing low accuracy. It may continuously read the same temperature with high precision, but that temperature does not reflect reality. As a friend pointed out, a broken clock is 100% accurate twice a day! A chemical assay may be highly reproducible and have great precision, but the compound being measured is not the chemical one thinks it is. Calibration of scientific instruments is crucial to ensure accuracy. Some instruments may be inherently imprecise because of limitations in their technology. These will be discussed in detail below.

The Instruments of Weighing: Balances and Scales

Now that we know atoms have mass, how do we go about weighing them? Scales in the form of simple balances (Fig. 6.4), have existed since ancient times. A measurable weight results from the interaction of a mass with an acceleration or gravity. *Einstein's Principle of Equivalence* states that gravity is indistinguishable from acceleration. *Newton's Second Law* states that weight equals mass multiplied by acceleration. We will assume that we are weighing a substance on the surface of the earth where weight measured in different locations will consistently reflect mass because gravity is constant. The difference in the gravitational field between Death Valley in California and the summit of Mt. Everest is approximately 0.03–0.04%, a small enough value that is ignored because most research is conducted close to sea level and thus relatively constant. Scales used on Earth do not work in zero-gravity space. Astronauts must measure mass using inertia in an accelerating field, which is equivalent to gravity. This is accomplished using a spring-loaded device called the Space Linear Acceleration Mass Measurement Device (SLAMMD).

Fig. 6.4 Balance scale used for weighing objects

Finally, if the compound is a gas, it can be weighed on a special purpose gas balance, weighed in sealed containers tared to the evacuated container, or the gas mass can be calculated using the ideal gas law and measuring pressure in a gas syringe. Liquids are measured directly by pouring into a container that has been tared and then weighed, or by measuring its volume, and calculating weight by correcting for the liquid's density. A zero weight, reflecting an empty vessel, would be limited by the sensitivity of the scale being used.

Chemists weigh chemicals using different types of scales. Figure 6.4 shows a primitive balance where weight of the sample in one pan is determined by adding known weights to the other pan until the scale is balanced. This basic design has continued to be used up to the present with improved methods to determine balance. A good scale can measure down to 100 mg and an electronic balance scale can reach 1 mg. Other scales use springs and tension rods, calibrated to known weights, to determine the weight of the sample. The precision of the scale is thus a function of the size of the weights used to calibrate the mechanical mechanism. *A general rule of thumb is that a scale can measure accurately up to one half of its smallest graduation.* The minimum weight of a substance that can be weighed on a balance within an acceptable percentage error (usually 5%) is termed the *minimum weighable quantity* (MWQ). Thus, for practical purposes, anything less than MWQ would be considered zero. But, as we will see throughout the history of chemistry, if a more sensitive scale comes along, like those to be described, MWQ will drop as will the operational definition of zero. The MWQ is very similar to the limit of quantification in a chemical assay to be discussed below.

Modern scales use electronic load cells to measure weight. Load cells are force transducers that convert force, such as a strained metal transducer beam, into an electrical signal that can be measured. Depending on the nature of the transducer, resistance or capacitance can be measured. As technology has improved, scales have become increasingly sensitive using magnetic fields or even the deformation of a crystal changing its electrical resistance due to the piezoelectric effect, allowing measurement of ever smaller quantities. It should be noted that some instruments used by physicists can weigh individual protein molecules and even protons, with the determination of mass being based on complex equations in quantum mechanics. This approach is not used in modern analytical chemistry laboratories.

The most sensitive laboratory scales in common use today are analytical scales that can measure down to about 0.01 milligrams. To obtain accurate measurements, all vibrations must be dampened, and the weighing pan isolated from drafts. Contaminants, such as dust or oil, must be eliminated. A modern laboratory scale is shown below in Fig. 6.5. These scales must also be calibrated before use. The accuracy of any measurement is dependent upon how the physical mechanism that transduces weight to a measurable response is correlated to an actual weight during calibration, this is also a function of the accuracy of the calibration weights. To ensure reliable estimates, replication is *de rigueur*. As will be discussed shortly, calibration and replication are essential parts of all analytical determinations.

Fig. 6.5 Modern analytical balance

Analytical Chemistry

The primary method used by scientists to estimate the amount of chemical present in a sample or its concentration in a solution, is to use instruments that detect a chemical based on its chemical or physical properties. Such techniques include light absorbance or fluorescence after the chemical has been extracted from its sample and put in the form amenable to instrumental analysis. Samples must often be 'cleaned up' to remove potential substances that would interfere with detection or chemically react with a compound that would allow detection. Detection and quantification are two different values for describing sensitivity of an analytical technique. The quality and care of how each of these broad steps outlined in Fig. 6.6 are performed contribute to the sensitivity, accuracy, and precision of the assay. Errors or carelessness in any one step will affect the results.

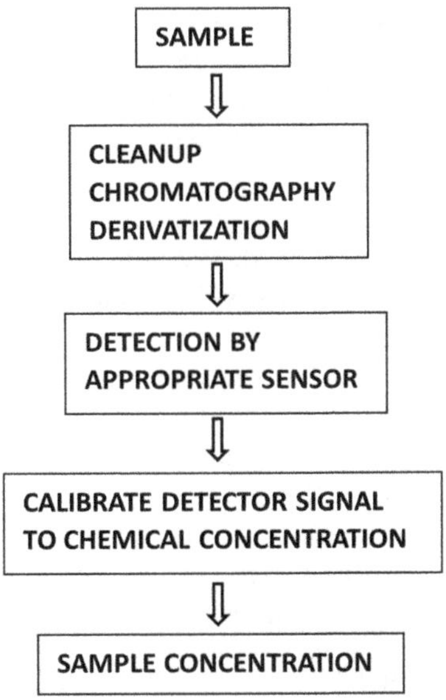

Fig. 6.6 Flow chart of a typical analytical analysis

The following are the chemical or physical properties used to detect and quantitate chemicals.

- Spectroscopy is the technique that measures the interaction of a molecule with electromagnetic radiation. These include atomic absorption and emission, infrared, visible, or ultraviolet light, X-ray, fluorescence, Raman, nuclear magnetic resonance, and photoemission spectroscopy as well as polarization interferometry. More specifically, for absorption and emission spectroscopy, a chromophoric group that is present in the molecule (a molecular moiety that absorbs light at a specific wavelength and/or emits at a different defined wavelength or color) is taken as a handle for its quantification. If a chromophoric group is absent, a colorimetric technique is used where a specific reagent develops color for quantitative estimation. For example, commonly used pH indicators found in swimming pool tests strips estimate the presence of acidic and basic compounds, showing red with acids and blue with bases.

- Mass spectrometry (MS) measures the mass-to-charge ratio (m/z) of one or more components of molecules present in a sample. Mass spectrometers convert molecules into ions which are then manipulated by electric and magnetic fields to obtain a mass spectrum that is essentially a fingerprint of known molecules. These measurements can be used to calculate an exact molecular weight and identify the sample components as well as determine its mass. MS is one of the most sensitive analytical techniques currently available and can commonly measure concentrations down to nanogram *per liter (1 part per trillion—ppt)* and even picogram per liter *(1 part per quadrillion—ppq),* though routine assays have limits in the microgram per liter (*1 part per billion -ppb*) range.
- Derivatization is the process of chemically modifying the sample compounds to become visible to the detection techniques above. Derivatization usually involves the addition of some well-defined functional group that is reliably detected by the detector being used. The goals of derivatization vary, but include increased volatility, greater thermal stability, modified chromatographic properties, adding fluorescent moiety, greater ionization efficiency, favorable MS fragmentation properties, or a combination of these.
- Immunological techniques using antibodies specific for the study molecule are another method commonly used in screening assays. This is the technique used to detect COVID-19 viral particles in home test kits. The enzyme-linked immunosorbent assay (ELISA) is a common assay used under laboratory conditions. It's advantage compared to the above physical techniques is their specificity requiring little to no sample preparation because they are specific for the compound being analyzed. Although interference can occur from similar compounds, they are often used in medical and food safety situations because they are fast and do not require extremely expensive analytical equipment. Detection limits are usually in the parts per million to parts per billion range. Other immunological techniques include approaches such as western blots.

Sample preparation is critical to successful detection in most of the above methods, except the immunological approaches above, especially when analyzing complex matrices such as blood, tissues, and food products. Techniques include capillary electrophoresis, protein precipitation with centrifugation or filtration, solid-phase extraction, liquid-liquid extraction, and affinity enrichment followed by separation techniques such as gas chromatography (GC), high performance liquid chromatography (HPLC), and Ultra-Performance Liquid Chromatography (UPLC). The most common approaches routinely used in laboratories today are hybrid systems, including HPLC–MS, GC–MS, GC–infrared spectroscopy, and GC-FID (flame ionization detection); the selection often being a function of the physical properties of the analytes being detected.

The actual sensitivity of many of these techniques is dependent upon the molecular weight of the compound being analyzed. Simply put, analyzing simple molecules like iron or lead is different than large organic molecules and large molecular weight proteins. *This is where molarity rather than concentration becomes important as chemical reactions occur between individual molecules and the number of molecules in a sample becomes critical for biological interpretation.*

There are some early analytical methods that now are largely of historical interest. In the early 1950's, chemicals were often measured simply by the absorbance of light in a spectrometer without separation. Antimicrobials were measured using bioassays where concentration was correlated to the ability to inhibit microbial growth in culture media. They suffered from a lack of specificity and sensitivities of these older assays being in the ppm range. Use of these assays persisted well into the end of the last century. Older immunological-based assays used radiolabeled antibodies to quantitate binding (Radioimmunoassay—RIA). These have been replaced by ELISA. There are some approaches designed for specific compounds, such as some hormones, that bind strongly to specific receptors and can then be easily quantitated in assay platforms that have these receptors bound to detector or reporter molecules. MS-MS (tandem mass spectrometry), HPLC-MS and GC-MS have largely replaced these approaches today.

The above overview illustrates the complexity of modern analytical chemistry. The accuracy and precision of the assay is a function of all the elements in Fig. 6.6. Errors at each step are cumulative. The author has used many of these techniques over his career for detecting contaminating residues in meat and milk and monitoring drug concentrations in animals and *in vitro* systems and even a hormone receptor assay for detecting thyroid hormones in the blood of molting harbor seals. Sensitivity in what were once routine assays has increased from parts per million to parts per trillion. This level of sensitivity has in many cases exceeded its biological meaning.

Statistics Involved in Determining Limits of Detection and Limits of Quantitation—The Operational Definition of Zero

The problem of limit of detection, or limit of quantification, is encountered in all scientific disciplines. This explains the variety of definitions and the diversity of solutions developed to address the question. We will try and make sense of this relative to the analytical chemistry involved in detecting residues of chemicals in biological tissues and food.

A calibration curve is a linear regression equation (recall $y = mx + b$ in statistics chapter). In chemistry, "y" would be the detector signal or response produced by the instrument by which the assayed molecule is detected (*e.g.*, electrical-voltage, luminescence, energy, *etc.*). "*b*" is the value where the fitted line intercepts the ordinates axis. "*m*" is the slope of the line of the function relating the measured signal to the quantity of the chemical being determined. "*x*" is the concentration of the chemical determined from the signal (S). In addition to the standard concentrations used to fit the regression line, replicate samples are conducted to determine reproducibility, blank samples with no study chemical (*true zero concentration*) to determine instrument noise, and control samples not involved in determining the calibration curve to determine accuracy and precision of the assay. For most skilled analysts using properly calibrated and maintained equipment, R^2 is generally well above 0.95 and often 0.99. However, as I can attest to from serving as Editor of a veterinary pharmacology journal for some 25 years, some research has been submitted with R^2 values of 0.7 to 0.9.

Two key parameters that are used to characterize the sensitivity of the assay are the limit of quantification (LOQ) and the limit of detection (LOD). The *LOQ is the lowest concentration of a substance that can be measured with certainty using standard tests. The LOD is the concentration of the chemical that can reasonably be detected in a sample with an acceptable level of accuracy and precision.* Although these definitions appear precise, there has been considerable debate among scientists as to how to precisely determine these lower limits in practice. The *precision* of the assay is determined by running replicates of samples while the *accuracy* is determined by comparing the assay's predicted concentration to a control sample of known concentration also included in the assay. Regulations can specify the amount of calibration samples that must be used and the statistical criteria for determining LOQ and LOD. Regulations promulgated by agencies, such as the Environmental Protection Agency (EPA) for environmental analysis, the Food and Drug Administration (FDA) for food and drugs, and the United States Department of Agriculture (USDA) for foods, often define their own criteria for acceptable calibration curves. *There is no single definition in existence, and thus LOD and LOQ are not universal constants associated with a chemical.*

Before we present a few examples, we need to define a concept termed *instrumental noise* (N), which is random fluctuations in the output signal of an electronic analytical instrument due to fluctuations in electrical components, temperature changes, and other sources of random noise. The exact threshold used to decide when a signal significantly emerges above the continuously fluctuating background noise remains arbitrary, often is a matter of an organization's policy and remains a topic of debate among scientists, statisticians, and regulators depending on the stakes in different fields. With this background, the following is a sample of definitions of LOD and LOQ in use today.

- The LOQ is the limit at which the difference between two distinct signal values can be discerned with a reasonable certainty.
- The LOD is defined as the lowest quantity of the analyte substance that produces a signal at least three times the average noise level of the detector, termed the signal to noise ratio or $S/N > 3$. This is believed to be when the signal is statistically different from the background.
- The LOQ can be calculated using an appropriate standard or sample and may be obtained from the lowest calibration point on the calibration curve, excluding the blank.
- LOD is the lowest quantity of a substance that can be distinguished from the absence of that substance (*a blank value*) with a stated confidence level (generally 99%). The detection limit is estimated from the mean of the blank, the standard deviation of the blank, the slope (analytical sensitivity) of the calibration

plot, and a defined confidence factor (*e.g.*, $\cong 3$ being the most accepted value for this arbitrary value).

- The instrument detection limit (IDL) is the analyte concentration that is required to produce a signal greater than three times the standard deviation of the noise level. This may be practically measured by analyzing eight or more standards at the estimated IDL, then calculating the standard deviation from the measured concentrations of those standards.
- LOD is defined as the lowest concentration of an analyte in a sample that can be consistently detected with a stated probability (typically at 95% certainty)
- The LOD according to IUPAC (International Union of Pure and Applied Chemistry) is the smallest concentration, or the smallest absolute amount, of analyte that has a signal statistically significantly larger than the signal arising from the repeated measurements of a reagent blank.
- The EPA LOD is calculated by multiplying the standard deviation of results from 10 injections of low concentration samples by 3.
- The FDA LOD is the smallest measured concentration of an analyte from which it is possible to deduce the presence of the analyte in the test sample with acceptable certainty. There are several scientifically valid ways to determine LOD and any of these could be used as long as a scientific justification is provided for their use.
- The LOQ is a stated multiple of the limit of detection at a concentration of the determinant that can reasonably be determined with an acceptable level of accuracy and precision.
- LOQ is the lowest concentration of a chemical that can be reliably quantitated with a S/N ration of >10.
- The FDA LOQ is the smallest measured content of an analyte above which the determination can be made with the specified degree of accuracy and precision. As with the LOD definition, there are several scientifically valid ways to determine LOQ and any of these could be used as long as a scientific justification is provided for their use.
- The EPA LOQ is determined using water samples spiked at a low concentration to give a coefficient of variation of less than 15%. The quality control limit used for the coefficient of variation is 20%.
- Some workers and programs even add clarification to LOD by calling it the Lower Limit of Detection (LLOD).
- Chemists often use the term Below Quantifiable Limits (BQL) when a concentration cannot be measured with certainty.

With so many definitions, one can now appreciate the source of the confusion that arises about precisely defining the lower limits of a chemical analysis. This also allows some workers to *find a regulation that validates their data.* They are all using various statistical approaches at various levels of significance to determine the:

- Minimum amount of a substance where a concentration can be reliably measured (LOQ)
- Minimum concentration where a signal related to the presence of a chemical exceeds random background instrumental noise (LOD).

To see how this played out in a laboratory setting, I asked an analytical chemist at my former NCSU laboratory, James Yeatts and pharmacologist Dr. Ronald Baynes, to try changing the calibration curve for an established assay and see how much LOD and LOQ changed. They used a routine assay for the antimicrobial drug ampicillin in bovine plasma using UPLC-MS. The R^2 of the calibration curve was 0.99647. Repeated blank (water) samples were run to determine the signal from background noise. Their LOQ approach was the lowest concentration where accuracy and precision are in the acceptable range (accuracy being within 80 to120%; and precision being less than 20%). This matches the LOD when calculated based on an S/N of at least 3 and the LOQ is based on S/N of at least 10. The S/N ratios were determined by using the computer software program that came with the analytical system "*TargetLynx®*". The operator selects the noise range in the chromatogram from an area that contains no chromatographic peaks, in this case the range is just in front of the peak of interest, which is essentially the baseline. Figure 6.7 lists two sets of concentrations used to determine two calibration curves to see if changes in samples making up a calibration curve alter LOD and LOQ. Replicates were run at 0.001 and 0.005 ppm; 0.005 ppm would be the LOQ based on an RSD (%) of 5.7% and average accuracy of 95.7% (n = 5). The RSD (%) (relative standard deviation) for 0.001 ppm concentration was 34.1% with average accuracy of 73.1%(n = 5). As expected, the LOD and LOQ were different.

ORIGINAL STANDARD CURVE Concentrations in PPM	**ADDED STANDARD CURVE** Concentrations in PPM
0.001	0.0009
0.005	0.003
0.01	0.0075
0.05	0.03
0.1	0.075
0.5	0.3
1	0.75
-	1.1
LOQ = 0.005	LOQ = 0.003
LOD = 0.001	LOD = 0.0009

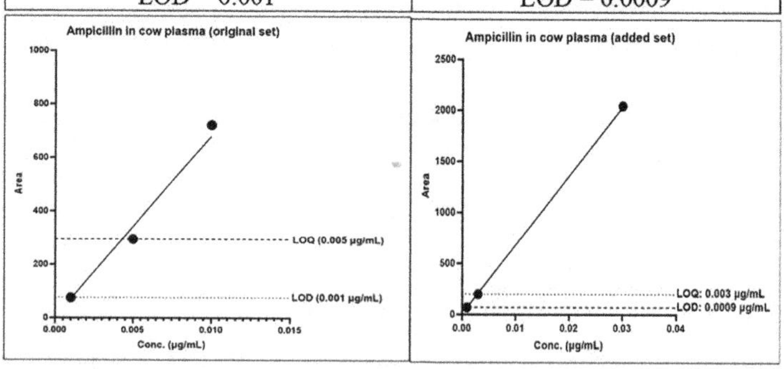

Fig. 6.7 Calibration curves showing how selection of standard concentration can affect LOD and LOQ. (Note that graphs only show lowest concentrations in the calibration curve to better visualize LOQ and LOD)

As can be clearly seen from this example and the sampling of the modern definitions above, *there is NO consistency in the definition of LOD and LOQ*. Many will converge on similar values in the hands of competent scientists using modern instrumentation properly maintained and calibrated, and having sufficient samples to determine replication. However, as discussed above, based on my editorial experience, this level of accuracy does not always occur.

Many regulators, including the FDA, leave the definition up to the scientists presenting their data for regulatory approval. The definitions above are guidelines and not necessarily standards. A *standard* is a document, developed through a consensus process, that clearly identifies specific, essential requirements for materials, methods, or practices for use in an unmodified form. A *standard* may, in addition, contain discretionary elements, which are clearly identified. In contrast, a *guideline* is a document, also developed through the consensus process, describing criteria for a general operating practice, procedure, or material for *voluntary* use. A guideline may be used as written or modified by the user to fit specific needs. Finally, when an LOD or LOQ are reported, the rules or guidelines used to define the reported values are often not even stated.

One approach is to define non-quantifiable concentrations as BQL using criteria selected by the chemists and report the result as BQL and not a number. This flags the value as not numerically quantifiable. However, other programs later scraping this data may just substitute zero, and the rules for defining one's BQL, are lost once the result is scraped.

Why do I think this issue is so important? As will be seen in later chapters, when so-called biologically safe concentrations are difficult to define, some regulations default to LOQ or LOD as thresholds that trigger legal responses. As we see, there are no absolute definitions of either LOQ or LOD. What happens if instrument sensitivity increases? This has continued to occur at increasing rates with recent technological advances, and LOD and LOQ for an assay will decrease yet regulations written decades ago remain unchanged. The limits detected today are well below those that existed when drugs were originally approved and may remain below safety thresholds but are now easily detected by modern assays. In this scenario various consumer groups and social media "experts" raise alarms when drugs and chemicals are *detected* in assays of food products, even if they are well below the safety thresholds. The opposite also happens, where more sensitive markers of biological effect lower the thresholds, sometimes below the ability to detect the compound. As will be discussed in detail in future chapters, these are crucial issues in modern society.

This is illustrated in Fig. 6.8. We assume that the concentration of a drug is X, estimated with an uncertainty of standard deviation Y. The estimate is thus a confidence interval of $X \pm Y$, with the lower limit often just being reported as X minus Y if a threshold is being defined. This could be 10 ± 1 ppm or 10 ± 2 ppt. Analytical techniques of varying decreasing sensitivities are represented by lines $A > B > C > D > E$. The most sensitive assay A has $X >> $ LOD, LOQ, and instrument noise. The concentration reported is real. In assay B, X is at the LOQ so the concentration reported would be the LOQ. However, in assay C, the sensitivity is lower still and there is a signal greater than LOD but not quantifiable so that it would be reported as LOD. Depending on how this is calculated, some workers have reported this as zero because it is at the sensitivity of the assay or as the average of LOD and LOQ. In assay D, a signal would be detected but not at a statistically significant level that distinguishes it from background noise, so it would be reported as zero or sometimes LOD. Finally, in the least sensitive assay E, any chemical present in the sample is well below the ability of the assay to see it and thus it would be reported as zero.

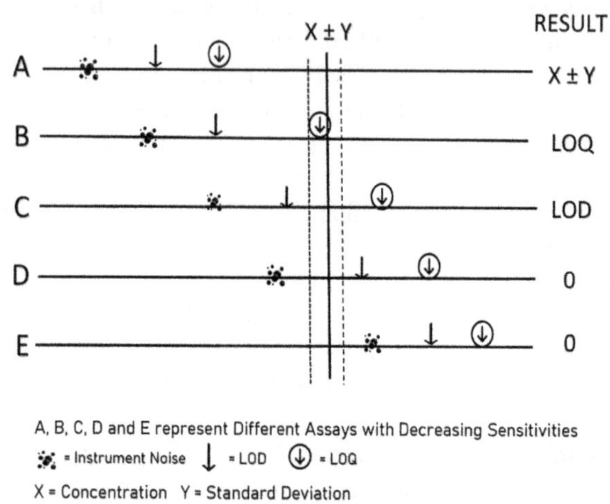

Fig. 6.8 Relationships between LOD and LOQ and the concentration reported out for the assay. In this illustration, the actual drug concentration is X with a SD of Y. Assay A is the most sensitive and can quantify the lowest concentration of residue present, in this case easily and accurately quantifying X. Assay E is least sensitive and cannot detect any signal that a residue is present at X concentration

These relationships clearly illustrate the dependence of a chemical's concentration upon the assay being employed to detect it. One could also view this scenario from a historical perspective where A is today, and E is a few decades back. This becomes important when regulatory limits are established for a compound using an older assay that can now be easily detected. We will return to this issue in future chapters.

Additional Concerns and Issues in Analytical Chemistry

The reader should appreciate that the discussion above is a broad overview of a very technical and rapidly advancing field. Analytical methods may be developed for very specific applications and then are adopted for uses never anticipated. In many cases, rapid diagnostic and screening assays are used that have higher error rates but are only meant to raise a flag as the presence of a potential contaminant or violative drug residue. These assays are meant to be validated using more precise analytical methods. Unfortunately, this validation step is sometimes skipped, and the screening results are presented as final results.

An area that is often neglected are the impacts of sampling protocols on the ultimate value of the concentration measured. There are numerous issues involved, including taking homogeneous and representative samples from a biological tissue, binding of the chemical to a substrate in the sample medium, and interference from similar substances. A partial list of these issues would include:

- Collection of a biopsy from a complex organ like muscle, liver, lung, or intestines where the biopsy taken may not be representative of the entire tissue. This is especially troublesome today when such small sample volumes are needed for analysis. The solution is to take larger samples and homogenize them to ensure the analytical sample is uniform and representative.
- The sample is at the injection site of the drug which is not representative of tissues throughout the body.
- The chemical of interest may not exist in a freely extractable form, as it could be covalently bound to a tissue protein, be locked inside a cell's organelles, or simply not be soluble in the solution used to extract the sample. The latter is an issue with general screening assays that purport to measure very water- and fat-soluble compounds in a single assay, yet such compounds may not be extractable using a single sample solvent.
- The compound itself is a stereoisomer where only one of the isomers are biologically active, yet the assay determines both.
- The chemical being assayed is not the parent chemical but a metabolite.
- The compound being assayed is not stable in the sample or extraction medium and may breakdown or precipitate, both processes removing compound from being analysed.
- The compound is not stable after being collected and may be required to be immediately frozen, stored under anaerobic conditions such as using nitrogen to prevent oxidation, or shielded from ultraviolet light which could degrade the chemical.
- There are often inaccuracies in measuring the volume of solutions used to extract a sample and to take a sample of the extract.

These experimental variations all add to overall assay variability. These sampling issues are dealt with using *spiked control* samples. This can be accomplished by adding a known amount of an analyte to a sample matrix before being processed and the results compared to the known amount of compound added to the matrix, with acceptable results often being 85 to 115% of known concentration. The use of such internal standards also helps when dealing with inaccuracies introduced during sample preparation. This procedure determines the accuracy of an assay. This approach has its issues for adding a chemical to a homogenate, which may not be the same as how the chemical naturally gets into the tissue. One can see that when running an assay, in many cases most samples are not unknowns, but rather are calibration samples and quality control samples such as spiked samples (accuracy), replicate controls (precision), and blanks (sensitivity). All of these add to uncertainty in the final value being determined.

Conclusion

This brief overview of basic chemistry and analytical chemistry provides the basis for further discussion about what zero truly means when applied to various disciplines. It is important to stress that no matter what the applications where zero has imputed significance, it is the analytical chemistry parameters that are providing the concentration to the sensitivity and accuracy it allows. We will refer often in this book to LOD and LOQ, so it is crucial that the limitations inherent to their determination be kept in mind.

Selected Readings

Holcapek M, and Brydwell W. H*andbook of Advanced Chromatography / Mass Spectrometry Techniques*. New York, Elsevier, 2017.

Housecroft C. *Inorganic Chemistry, 5th Ed.* New York, Pearson, 2018.

Hudson J. *The History of Chemistry.* London, Red Globe Press, 1992.

Kisch B. *Scales and Weights—A Historical Outline.* New Haven, Yale University Press, 1976.

Manz A, Dittrich PS, Pamme N, and Iossifidis D. *Bioanalytical Chemistry, 2nd Ed.* London, Imperial College Press, 2015.

Partington JR. *A Short History of Chemistry. 3rd Ed.* Garden City, Dover, 2011.

Robinson A. *The Story of Measurement.* London, Thames and Hudson, 2007.

Skoog D, West D, and Holler F. *Fundamentals of Analytical Chemistry, 10th Ed.*, Boston, Cengage, 2021.

United States Pharmacopeia (1225) *Validation of Compendial Procedures*. 2021. https://online.uspnf.com/uspnf/document/1_guid-e2c6f9e8-ea71-4b72-a7ba-76abd5e72964_4_en-us

Chapter 7
Biology and Pharmacology

> Biology is the study of the complex things in the universe. Physics is the study of the simple ones. (Richard Dawkins)

Our next task is to investigate the implications of zero and nothing in the realm of biology and pharmacology. The discussion grows more complex when living matter enters the picture. Life can be defined as *what distinguishes animals and plants from inorganic matter, including the capacity for movement, reproduction, growth, and development preceding death.* Living organisms have the ability to fight or export entropy, otherwise it would drive them to be in equilibrium with their environment, an intriguing definition of death. These are broad definitions, but they help define concepts such as the *absence of life* or *zero biological effect* after low-level chemical exposure.

Topics to be discussed in this chapter range from molecular to cellular biology to the anatomy and physiology of complex multicellular organisms. This overview will only include information sufficient to allow the reader to grasp the implications of zero in this field. This requires an understanding of the molecular interactions that define life and how molecules interact at the various scales of biological organization. This knowledge allows one to intelligently discuss issues of toxicology and food safety related to *what is a safe level of chemical exposure to humans.*

What are the molecular and cellular interactions that can occur between a chemical and biological structures? Two major processes that modulate chemical biological interactions occur at most levels of biological complexity:

1. *Gaining access to a receptor that initiates a biological effect.* This includes processes such as absorption, distribution, transport, metabolism, and elimination; processes often requiring passage through biological membranes.
2. *The chemical's interaction with a receptor or other biomolecules,* such as proteins and genes that initiate a biological response.

Figure Use Attributions
Figs. 7.1, 7.2, 7.3, 7.4, and 7.5: Riviere JE. *Comparative Pharmacokinetics: Principles, Techniques and Applications,2nd Ed*, Ames, IA, Wiley-Blackwell, 2011. c Jim Riviere
Figs. 7.6 and 7.8: Wikimedia – No restrictions
All other figures are either self-drawn, photographed or from author's own publications (attributed)

J. E. Riviere, *Zero – Much to Do About Nothing?*,
https://doi.org/10.1007/978-3-031-82998-7_7

The first process is dependent upon the concentration of chemicals in the exposure scenario. How does that chemical cross the membranes that create the fundamental structures that define cells and organisms, and how is it carried between and within these structures? A few basic physical chemical properties generally characterize these processes. The second process defines the mechanism of biological effect and for our purposes will largely be an adverse toxicologic effect.

When debates occur over what is the dose of a chemical that can cause harm to humans, commentators often focus on the molecular interactions that initiate an event, such as an interaction of a chemical with a specific receptor or a chemical interaction with DNA (deoxyribonucleic acid—the basic molecule of our genes). It's molecule to molecule or molecule to a molecular receptor. It is the result of such interactions on the function of an individual cell or organ, as well as the health of the entire organism, that is the ultimate endpoint. Most effects occur within individual cells. When serious damage occurs to critical cells, the organism itself may be adversely affected. These topics fall into the realm of pharmacology, toxicology, and pharmacokinetics; areas that define my professional career. For more in-depth details, textbooks listed in the selected readings section, or any general pharmacology or physiology text should be consulted.

This chapter will introduce the "characters" involved in "the play of life". The cast in this drama ranges from the invading molecule to their target receptors, cells, organs, organisms, and finally to the entire ensemble. Depending upon the script, a cell will have different roles to play. In this discussion, we will only introduce the characters and how they are scripted to interact with one another. In the following chapters, we will explore specific plays involving cancer, food safety, and other toxicological endpoints. The key to understanding the implication of zero and nothing is to realize that the cast of characters are limited and defined by evolution, with the interpretation of their actions dependent upon the play that is being performed. In the same way that a unique sound made from individual instruments is very different when playing in an orchestra and a new blended sound emerges, so do complex physiological functions emerge from the coordinated network of individual cells and organs.

Some perspective on cell death in a multicellular organism is needed to assess how serious a toxicological response can be. The average adult human loses between *50 and 70 billion cells* every day due to normal cell death and cell turnover. Cells undergo a process of programmed death that is referred to as *apoptosis*. In the time it took you to read that sentence, almost one million of your cells died in your body. *Cell death and turnover are normal physiological states.* Damaged cells from chemical exposure may die and are replaced by normal cells. In the skin, ultraviolet light exposure (*e.g.*, sunburn) is constantly killing cells. Additionally, all cells are not created equal. Damage to reproductive cells can produce harm to the genes that in turn get transmitted to the offspring without harm to the parent's cells. As discussed in the body of this book, it is when the rate of cell death exceeds replacement, or damaged cells escape the multiple safeguards that the body has to weed out potential problem cells, that toxicity will occur.

A Primer on the Organization of Multicellular Organisms

One approach to frame this discussion is to look at the body from the perspective of where a molecule goes to interact with a cell in the body, which is a roadmap for transit through the body. The 1970 movie *Fantastic Voyage* was such a journey. Let's start this discussion at the level of the organism and then proceed to discuss interactions within cells. Figure 7.1 is a schematic of the routes that a chemical may take to interact with and distribute throughout an organism. The primary routes are:

- Topical exposure to interact with a multicellular organism via the skin (TOP)
- Oral exposure by the mouth through the gastrointestinal tract (PO)
- Inhalation through breathing a compound in via the mouth or nose (IN)
- Injection into the body by a needle which is primarily used for drugs. Injections can occur:
  - Subcutaneous under the skin (SC) or into the dermis (ID)
  - Intramuscular (IM)
  - Directly into the vascular system (IV).

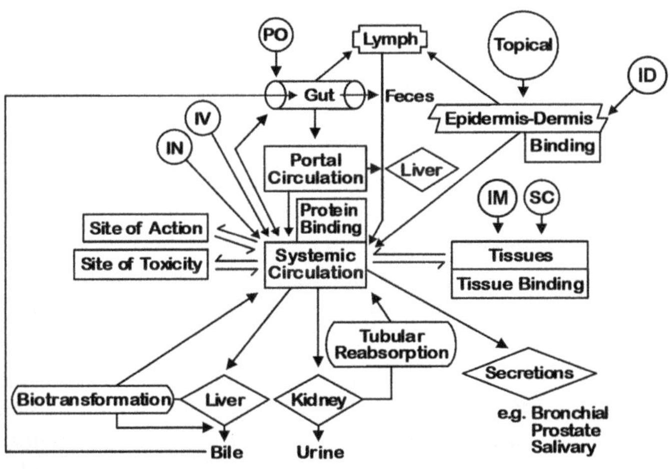

Fig. 7.1 Absorption, distribution, metabolism and elimination (ADME) processes

All these routes of absorption require a significant number of molecules survive the journey to interact with their target cell. Compounds reach their target tissues after entering the systemic circulation in a process termed biodistribution. If the compound is distributed through the systemic vascular circulation, it may reach organs such as the liver and kidney which can metabolize them to inactive compounds. For drugs and chemicals, this is called *biotransformation*. Also, the kidney and liver may eliminate unchanged or metabolized compounds from the body into urine and bile, respectively. In some cases, metabolism can activate a harmless chemical into an active one thereby causing toxicity in the excretory organ or other cells. This is an important point when we discuss direct and indirect carcinogens in Chapter Nine. These critical processes of absorption, distribution, metabolism, and elimination are termed the ADME processes. Note in Fig. 7.1 that some tissues are marked as beneficial (site of action) targets while others are sites of adverse effects (site of toxicity). The physiology of getting a chemical to these sites is identical, using our play analogy, the actors are the same, but the script is different once they arrive. This distinction is often dose-dependent for chemicals and drugs.

All cells lining tubular passages in complex multicellular organisms generally excrete lubricants into their lumens. Compounds can be excreted by sweat ducts and sebaceous glands in the skin, bronchial glands in the respiratory system, the lacrimal glands in the eye, the sebum glands into the ear canal, the salivary glands in the oral cavity, the mammary glands into milk, the prostate into the urethra, and in individual gastrointestinal tract lining cells into its lumen.

Compounds that are water soluble and easily dissolved in plasma or extracellular fluid are readily eliminated from the body primarily by the kidney. If a lipophilic (fat soluble) compound is absorbed, the liver generally will metabolize it to a hydrophilic (water-soluble) metabolite, which then can be eliminated. If a lipophilic compound can't be metabolized easily, it may persist in the body for long periods. The body has been honed during evolution to possess defense systems for protection from dangerous chemicals that enter the body in sources such as the toxic plants our ancestors consumed. When compounds are absorbed in the intestines, the blood supply is carried by special vessels called the *portal veins* directly to the liver before they reach the rest of the body. Here, they may undergo so-called *first-pass metabolism* where they are transformed to compounds that are more easily excreted by the kidney. In some cases, for very specific chemicals, the metabolites themselves may be toxic. In other cases, the lipophilic compounds are so fat soluble that they become sequestered into tissue as described below which offers them sanctuary from further metabolism and rapid elimination.

It's instructive to see how complex the journey is through just the digestive tract (Fig. 7.2). There are numerous points where compounds can interact with cells lining the digestive tract starting in the oral cavity of the mouth, continuing down the esophagus, and into the stomach, small and large intestines, and finally excretion out of the body as feces. At every step, the digestive tract has defenses to isolate potentially toxic chemicals, ranging from binding to mucous in the mouth, to destruction by the strong digestive acids in the stomach, enzymatic digestion in the small intestines, bacterial metabolism in the colon, or expulsion of very irritating substances by spitting or vomiting and then finally by excretion through the colon. For example, most proteins are digested in the stomach because they normally are eaten for sustenance with gastric acid hydrolyzing the proteins to release amino acids essential for nutrition. If a protein drug is swallowed, it too will generally be digested and have no systemic effects. This is the reason why there are no oral insulin pills available for diabetes. The small intestines are optimized for absorbing gastric-digested nutrients from food. They accomplish this by expanding their absorptive surface area; at the macroscopic level by folding into the lumen and at the microscopic level by digestive cells having microvilli on their surfaces. The microvilli increase the surface area some 600-fold over a simple tube.

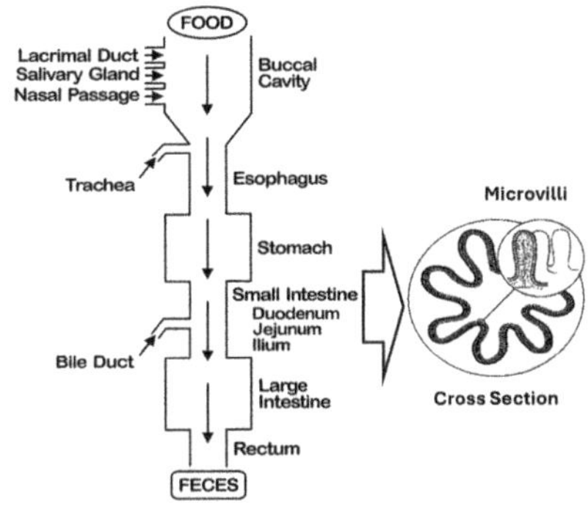

Fig. 7.2 Plumbing of the gastrointestinal tract

A similar tortuous journey exists for compounds with direct contact with the skin. The skin has evolved to serve as a barrier to keep foreign chemicals from getting absorbed and from letting water evaporate through skin. This is illustrated in Fig. 7.3. The outermost layers of the stratum corneum appear like a brick-and-mortar wall, with the transmission electron micrograph insert depicting them as flat shingle-like structures. These are stacks of dead protein rich cells that are excellent barriers to chemical penetration. Compounds penetrate primarily through the intercellular lipid mortar (A) requiring them to be somewhat fat soluble to cross the skin barrier. Smaller amounts are absorbed transcellular (B), through hair follicle shafts (C), or via sweat ducts (D). A magnified view of the intercellular path is depicted in the inserts below.

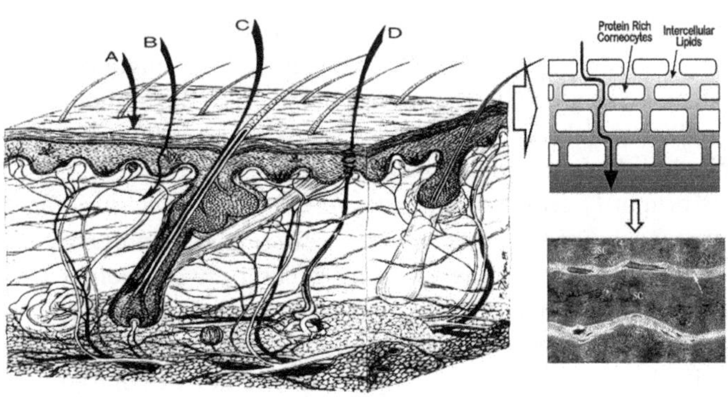

Fig. 7.3 Pathway for the journey through the skin

Membranes and Transport Conduits

A different perspective is to increase the magnification and look at the structure of the microscopic membranes that a compound must pass through to get to the surface of or inside a cell where it could trigger a biological effect. Membranes are a fundamental construct of biology that allowed cells to evolve some four billion years ago. Membranes separate "in from out", a unicellular organism from the water or air, a cell from the blood around it, or a nucleus from the intracellular environment. For chemicals to get into an organism or a cell or a subcellular organelle, different membranes with different structures must be traversed. Figure 7.4—A is a simplified schematic illustrating some potential routes of compound movement across a generalized cellular membrane.

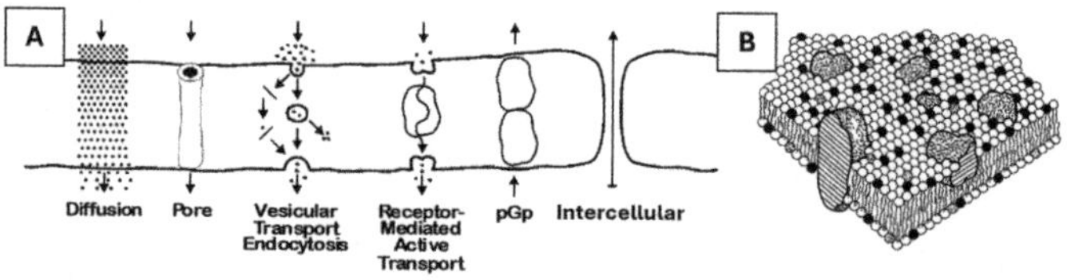

Fig. 7.4 Cellular membrane function and structures involved when chemicals cross a cellular membrane

This schematic illustrates the functional elements of transporters, receptors, vesicular transport systems, pores, and other biomolecular functions that define a cell's function and its role in an organism's physiology. Numerous other protein complexes with different functions may be present. The actual structure of a biological membrane consists of a bilayer lipid membrane in which large protein structures are embedded as depicted in insert B above. This model was proposed in 1972 as the Singer–Nicolson Fluid–Mosaic Membrane Model which remains conceptually correct today. These membrane-associated proteins are the functional units of the membrane. A transporter protein or pore would completely traverse the membrane allowing for transcellular passage, while a receptor might only be located on one side. This model is an obvious oversimplification of the tremendous complexity seen in biological membranes, but for our purposes it nicely illustrates the level where chemical-biological interactions occur.

Chemicals primarily cross membranes either by *diffusion, molecular transporters,* or within *specialized transmembrane vesicles.* Transporters can vary from being ultra-specific to just honing on some general physical chemical properties, such as the molecule simply being a weak acid or a weak base. Other transmembrane enzymatic systems transport ions out of the cell to maintain a cell's homeostasis. Each cell type has evolved specific assemblages of transporters in their membranes related to that cell's function in an organism. Kidney and liver cells have abundant transporters for chemical secretion. Nerve cells have membranes both suited to the ion transmission involved in generating a neural pulse as well as receptors at the junctions between neurons (synapses) to continue transmission between different neurons. Immune cells, like lymphocytes, macrophages, and neutrophils, have surface receptors searching for foreign compounds, viruses, or bacteria, as well as active vesicular transport systems to ingest and destroy foreign matter. However, for a compound to get to a cell in a multicellular organism, membranes must be traversed at some point in their journey.

Every biological compartment in a complex multicellular organism is thus defined by limiting membranes, including the skin, gastrointestinal tract, endothelial cells lining the capillaries, and then at the cellular level, membranes that define the cell and its intracellular organelles. This journey across membranes on the microscopic level is depicted in Fig. 7.5.

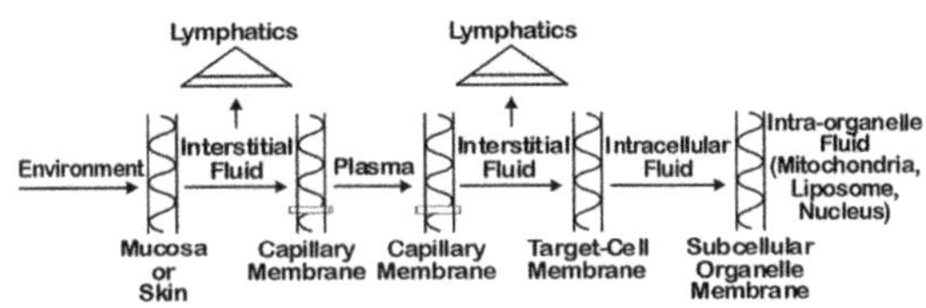

Fig. 7.5 Journey of a chemical from outside the body through membranes and body fluids before reaching target cell

These membranes have very different structures. As was seen in Fig. 7.3, the outer skin membrane is composed of multiple cell layers, making the skin an effective barrier to keep things out and in fact keep water in! Erosion of the skin barrier as seen with severe burns can result in evaporative water loss and severe dehydration. The author views the gastrointestinal tract as a tube passing through the body and can be considered an area external to it. Skin and the gastrointestinal as well as the respiratory tracts have immune cells located in the epithelium at this interface with the outside world. This is usually the site of initiating an immune response, such as after receiving a nasal vaccination or an allergic response after exposure to an allergen such as poison ivy. In contrast to skin, intestinal cells have very specialized transport systems that allow absorption of compounds such as nutrients, and excretion of other substances such as gastric acid. Excretions from the lacrimal and salivary glands as well as bronchial and hepatic biliary tract excretions add compounds back into the gastrointestinal tract. Inhaled compounds generally are gases or vapors that reach the very thin capillary membranes surrounding the lung alveoli, where they are transported by the vascular system. The upper respiratory system removes most particulate matter and returns it to the digestive system. The body evolved over eons to a complex organism with built-in safeguards for exposure to adverse substances.

The best example of these protective systems is the *blood-brain barrier* (BBB) which prevents many compounds from entering the brain. The transmembrane enzyme transporter p-glycoprotein (pGp in Fig. 7.4), which has been called *the guardian of the brain,* actively pumps harmful chemicals out of the brain. This is necessary because brain neurons are exquisitely sensitive to even small concentrations of foreign chemicals. By contrast, organisms lacking a BBB, (*e.g.*, insects, parasites) are susceptible to antiparasitic drugs (*e.g.*, ivermectin) at concentrations that do not harm most mammals.

Transport within the body is primarily through the blood vascular system although transport can also occur through the lymphatic system with delivery to the lymph organs. These vessels have walls composed of cells and their membranes. Compounds must either be dissolved in the aqueous medium (*e.g.*, plasma) or be bound to a host of different blood proteins depending on the water or lipid solubility of the chemical being examined. Many lipid (*e.g.*, fatty or hydrophobic) chemicals are only transported in the blood via nonspecific binding to proteins. In contrast, the interaction between a chemical and a protein can also be more specific, examples being hormones and compounds such as cholesterol (think low density lipoproteins—LDL *versus* high density lipoproteins—HDL in your clinical lab report). In some specialized areas, such as the nasal epithelium, some chemicals can enter the body through olfactory neurons, bypassing the BBB, and deliver compounds directly to the brain *via* neuronal transport. It's a complex system that prevents broad nonspecific statements from being made.

Physical Chemical Processes Involved in Chemical Movement

Knowing the landscape of biological systems that must be traversed by a chemical to reach its target site, what are the physical chemical properties and processes by which a molecule distributes throughout a system? In most of the whole-body pathways presented above, two simple physical processes, *diffusion* and *dilution,* are critical. Both require that significant concentrations of compound be present to distribute chemicals throughout the body and into target cells. *There must be sufficient chemicals to retain a high enough concentration gradient for molecules to both cross membranes or bind to and activate a receptor.*

Diffusion is an extremely important concept throughout biology, and especially so when we are trying to define when a concentration of a chemical is below the level required for a biological effect to occur. Briefly, the rate of diffusion of a compound across a membrane is directly proportional to its concentration gradient across the membrane (ΔX), the chemical's lipid solubility measured as the lipid:water partition coefficient (P), the diffusion coefficient (D), which is dependent upon molecular size, conformation and solubility in the membrane milieu, and membrane thickness or diffusional path length (h). This can be summarized by *Fick's Law of Diffusion*:

$$Rate\ of\ diffusion\ (mass\ /\ time) = \frac{D(\cancel{cm}\ /\ time)\cdot P}{h(\cancel{cm})}(\Delta X)(mass)$$

The term *(D•P/h)* is referred to as the *permeability coefficient* K_p, which is the determinant parameter of whether a chemical can easily cross a membrane. K_p is dependent upon the physical properties of the chemical and the specific membrane being crossed. We have previously encountered rate equations in the mathematics chapter. When used in modeling, they are written as a differential equation *(dx/dt)* and evaluated at an instantaneous rate approaching a zero-time interval. In membrane transfer studies, the total flux of drug across a membrane is dependent upon the area of membrane exposed; with the rate often expressed per cm^2. This is the mechanism by which microvilli enhanced the surface area and increased absorption in the intestines. A similar increase in absorptive surface area occurs in the alveoli of the lungs (Fig. 7.6).

Fig. 7.6 Adolf Eugen Fick—Author of Fick's Law

This concentration gradient is dependent upon the number of molecules capable of diffusing across a membrane. The actual movement across a barrier is a function of D, which is heavily dependent upon molecular weight. Simply put, a smaller molecule results in a larger value of D, making it easier to diffuse through most membranes. I mention this only to stress that detection of a large molecular weight molecule such as a protein present at very low concentrations may not be a toxicological hazard compared to the same mass of small lipophilic molecules.

For certain compounds, this scenario is complicated by whether or not the compound is charged which decreases the ability of a molecule to cross a membrane. Charged anions (negative charge) and cations (positive charge) cannot cross lipid membranes and are often transported by enzymatic transporters mentioned earlier. Also, a chemical's charge is dependent upon the pH of the environment it is in and to where it is going on the other side of the membrane. This phenomenon also determines where drugs are absorbed in the gastrointestinal tract. When the pH is different on two sides of a membrane, accumulation may result where the now-charged chemical in its new pH environment cannot return across the membrane. This mechanism, called *pH partitioning* or *ion-trapping*, is often used to target antibiotics to specific tissues where the local pH promotes drug accumulation.

If a molecule is a substrate of a specific molecular transporter system, such as the acid and base transport systems above, then constant amounts of chemicals may move across the membrane and not be dependent upon the concentration gradient, ΔX. In this case, these systems are very efficient at transporting very low concentrations of chemicals, but at very high concentrations, they can become saturated. Note there always will be some membranes (*e.g.*, capillaries) that need to be crossed by diffusion or crossed to get to the site of transport into a cell.

Dilution is another particularly important factor because once a chemical enters the systemic circulation; it will dissolve in the plasma medium of the blood and be diluted at minimum into the volume of the blood vascular system which is ≅ 7–8% of total body weight. This dilution decreases the concentration gradient driving the compound across the membrane. This percentage of body weight distribution increases if the compound can diffuse across the capillaries into extra-cellular fluid (≅20%) and is much larger if it can distribute throughout the body (≅60%) by getting access to the intracellular space. The extent of distribution through the body is quantitated by parameters such as the volume of distribution.

Similar to this dilution for hydrophilic water-soluble drugs, very lipid soluble or lipophilic molecules with high P, may accumulate in fat cells and adipose tissues. Recall that lipophilic chemicals can only be transported bound to proteins in blood. What generally happens is that upon passage through the vascular system they are extracted by the liver to be transformed into more water-soluble compounds for easy excretion, yet they distribute to adipose tissue and other fatty organs such as the mammary gland. Then they then can accumulate in these tissues upon repeated exposure. Paradoxically, this removal from the circulatory system protects the compound from being metabolized and reaching other body tissues. They may slowly accumulate until their equilibrium concentrations cause systemic toxicity. This is seen with increased adverse reactions to some very lipophilic drugs (*e.g.,* benzodiazepines) in older obese patients. Lipophilic contaminant chemicals, that can easily cross membranes and are prone to such accumulation, are referred to as persistent or forever chemicals.

Similar to lipophilic compounds being favorably attracted to adipose tissue and in some cases the fatty tissues of the nervous system, other compounds preferably accumulate in different tissues depending on their physical chemical properties. Some compounds sequester into bone or bind to specific proteins in tissues based on their charge, while others accumulate in subcellular compartments within cells *via* binding or ion-trapping. Other, like aminoglycoside antimicrobial drugs, accumulate in the kidney and potentiate renal toxicity. This has implications for interpreting what a very low chemical concentration means, because it is very chemical and environment dependent. Knowing there is chemical exposure but finding low to non-detectable concentrations in the blood (*e.g.,* analytically below LOD, hence zero) may indicate no exposure for certain compounds or for others may mask the fact that significant compound may be present in the body, but is diluted or more likely sequestered, in fatty tissues.

Scientists measure the tendency of compounds to accumulate in tissues by calculating the ratio of concentrations in blood *versus* tissue using the partition coefficients (P) defined above. This is practically estimated in the laboratory by measuring the compound's octanol / water partition coefficient, termed $K_{ow,}$ under constant and defined (generally neutral pH) experimental conditions. Octanol is a lipophilic hydrocarbon that mirrors the properties of lipids found in many membranes and fatty tissues. Because these partition coefficients are often very large, this parameter is reported as log K_{ow}. Highly lipophilic compounds include many environmental contaminants such as DDT and PCBs as well as PFAS today. Table 7.1 lists several chemicals we will encounter throughout this book. Low exposure causes tissue accumulations that persist for years. Recall that some compounds accumulate into specific tissues because of pH partitioning or strong tissue binding.

Table 7.1 Examples of chemical partition coefficients

Compound	Log K_{ow}
Aniline	0.90
O -Toluidine	1.29
Benzaldehyde	1.48
Acetophenone	1.58
Anisole	2.11
Benzene	2.13
Toluene	2.69
Styrene	2.95
Benzoic acid	1.85
Phenylacetic acid	1.30
Nitrobenzene	1.85
Ethylbenzene	3.15
n-Propylbenzene	3.68
1,3,5-Trimethylbenzene	3.42
t-Butylbenzene	4.11
Fluorobenzene	2.27
Chlorobenzene	2.84
Bromobenzene	2.99
Iodobenzene	3.25
Dichlorobenzene	3.38
1,2,4-Trichlorobenzene	4.02
Biphenyl	4.09
2-PCB	4.54
2,4,4'-PCB	5.62
Hexachlorobenzene	5.50
DDT	6.36

Data from Chiou and Block (1986)

Chemicals that are foreign to the body may interact directly with the protein immunoglobulins and other immune cells, inducing either an allergic reaction or an immune response. Immune cells are located within the vascular and lymphatic systems, as well as in cells that interface with the outside environment in skin, the gastrointestinal tract, and respiratory system. This important interaction, which does in fact respond to very low exposure concentrations, will be discussed in the next chapter.

Cellular Biology and Physiology

The above discussion has provided some basic descriptions of the components of membranes and organisms that govern translocation of chemicals in biological systems. The scale of biological entities ranges from viruses, bacteria, cells, and multicellular organisms to entire ecosystems. The focus of our discussions though is on the cells and structures that comprise the multicellular animal we know as *Homo sapiens*. The fundamental structural unit that is important on both microscopic and macroscopic levels is the eucaryotic cell depicted in Fig. 7.7. Eucaryotic cells are those found in multicellular organisms in contrast to cells of bacteria, fungi, plant cells, and the other organisms that compose our biosphere. Our focus will only be on mammalian eukaryotic cells.

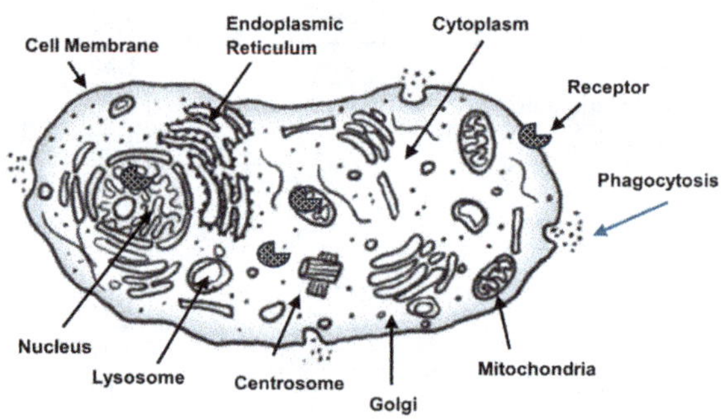

Fig. 7.7 Structural components of a eukaryotic cell

The major structures in the cell, including the chromosome- containing nucleus, mitochondria, and lysosomes are defined by their delimiting membranes, again barriers to movement of molecules. For the purposes of our discussion, these various compartments may be involved in the transport of chemicals into and within the cell or be the location where a toxicological reaction takes place. Thus, when we talk about a chemical binding to a receptor (●), that receptor is a membrane embedded protein complex located on the surface of a cell, within the nucleus or another intracellular organelle. These distinctions are important as location has a role in determining the dose required to have an effect as well as defining what the biological response will be.

The *endoplasmic reticulum* is a large organelle that is the backbone of a cell's transportation system. It is composed of a sprawling network of sacs and tubules that extend from the nuclear membrane throughout the cytoplasm through which proteins and other molecules can move. It is the cell's vascular system. It is a dynamic structure that serves many house-keeping roles in the cell, including storage of calcium, protein synthesis, and lipid metabolism. These diverse functions are performed in distinct domains, consisting of tubules, sheets and the nuclear envelope. The tubules and sacs all interconnect, allowing the endoplasmic reticulum membrane to form a continuous sheet enclosing a single internal space that is the site of production of all the transmembrane proteins and lipids for the cell.

The *Golgi body or Golgi apparatus* is a cellular organelle that helps process and package lipid and protein molecules, especially those proteins destined to be excreted from the cell. As a part of the endomembrane system in the cytoplasm, it packages proteins into membrane-bound vesicles inside the cell before the vesicles are sent to their destination.

The *centrosome* is a membrane-free organelle that serves as the main microtubule organizing center in animal cells. The microtubules provide the skeletal structure to the cell and regulate cell-cycle progression. Through this ability to organize microtubules, they are involved in cell polarity and cell division.

Lysosomes are organelles that contain enzymes that break down biological polymers. They are found in all animal cells but are most numerous in disease-fighting cells like macrophages and neutrophils. They have a lower pH than the rest of the cell and can accumulate compounds due to ion-trapping. During endocytosis, the cell membrane undergoes an invagination and forms a phagosome, which upon further processing fuses with a lysosome resulting in the ingested content being packaged within lysosomes and enzymatically digested. Endocytosis involves ingesting large particles, like bacteria, and is termed phagocytosis, while pinocytosis ingests smaller particles and liquids. Particles and chemicals entering the cell *via* this pathway usually do not cause any further biological effects. Some nanoparticles and some highly charged molecules can occasionally overwhelm the lysosomal digestive enzymes, persist, and disrupt cellular functions. Certain genetic conditions referred to as lysosomal storage diseases are characterized by dysfunctional lysosomal enzymes and cause serious damage to cells.

Mitochondria are membrane-bound cell organelles that generate most of the chemical energy via electron transport needed to power the cell's biochemical reactions. Chemical energy produced by mitochondria is stored in adenosine triphosphate (ATP) molecules that drive many cellular enzymatic processes. Because of their high oxygen metabolism, mitochondria are prone to oxidative injury. Some toxins, including specific non-steroidal anti-inflammatory, anti-retroviral and anti-diabetic drugs, the anti-cancer drug doxorubicin, as well as the antibiotic chloramphenicol are mitochondrial toxins.

In biology, the *nucleus* is a membrane-bound structure within a cell that contains chromosomes. The nucleus is the largest organelle in most cells and is responsible for storing genetic information, synthesizing, transcribing, and processing DNA and ribonucleic acid (RNA), and aids in coordinating cellular architecture. The nucleus is encased by a membrane with pores that allow certain molecules to pass into and out of the nucleus. Within the nucleus is the *nucleolus.* The nucleolus is a spherical structure whose primary function is to produce and assemble the cell's ribosomes, the structures used to transcribe RNA to manufacture proteins. The nucleus is the site of action of mutagenic chemicals, which are potent agents that act at low concentrations; and will be important in our discussion of carcinogenic and mutagenic toxins. A genotoxin is a chemical that causes DNA and then chromosomal damage. When damage occurs in a reproductive germ cell, it is termed a *mutagen* and has the potential to cause a heritable altered trait. If DNA damage occurs in a somatic cell, the cell type not involved in reproduction (most of the cells in the human body), malignant transformation of the cell may occur.

The reader must appreciate that the above overview is by necessity a generalization of complex systems. As we discuss the definition of the limits of biological effects, an assessment of their potential to occur is dependent upon what and where the target biomolecule or receptor is located both within the cell and the organism.

Chemical Interactions with Biomolecules

Now that we have defined the cellular targets and the organismic physiology that controls delivery and distribution of potentially toxic molecules, then we will discuss the interactions of a potential toxicant with receptors and other biomolecular interactions.

As introduced in the physics chapter, forces that govern molecular interactions include ionic attraction, hydrogen bonding, van der Waals forces, and other hydrophobic and hydrophilic interactions. Some interactions are nonspecific and do not initiate biological responses and may just result in tissue binding. In some cases, a molecule will directly interact with a non-protein target, such as when DNA strands combine to form a double helix or when DNA is transcribed from RNA. Some chemicals dissolve in membranes changing the way their embedded protein's function.

In other cases, the interaction is very specific and dependent upon the three-dimensional conformation of the molecule and its receptor, conceptualized as the so-called lock and key interaction analogy in Fig. 7.8.

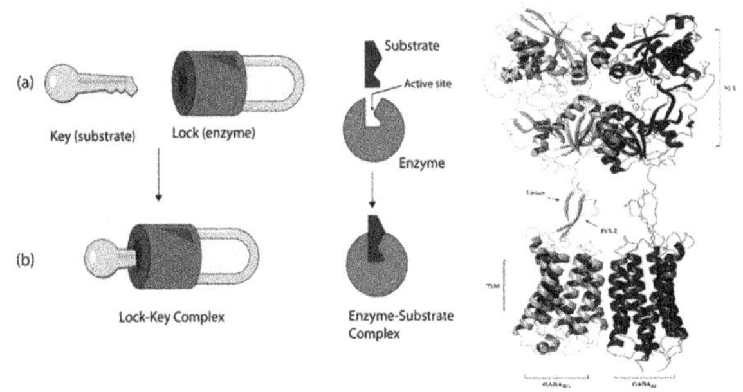

Fig. 7.8 Drug-receptor interaction. The left insert illustrates an oversimplified cartoon view where (**a**) a substrate approaches its target and when bound (**b**), activates the enzyme or opens the lock. The right insert approaches reality where protein strands making up the GABA receptor surround the target molecule

Recent research has shown this interaction is far more complex, with multiple points and strengths of intermolecular contact, but the concept holds well for our purposes since once a molecule fully engages with the receptor, an action happens. For specific receptors, binding results in interaction that triggers a biological effect. The mechanisms by which an effect occurs are numerous and include initiation of enzymatic activity, opening of ionic pores, release of ions or messenger molecules into the cell and translocation of molecules across the membrane. It is these specific interactions that often can occur at very low concentrations compared to the other non-specific mechanisms that require larger concentrations. Most receptors exist to regulate the basic metabolic functions of a cell or to allow transmission of signals across the organism. Hormone receptors are extremely sensitive and allow precise regulation of physiological functions at very low concentrations of hormones. The functions of specific receptor activation will be discussed in later chapters when specific chemicals are discussed.

The Dose Response (D-R) Relationship

A chemical - receptor interaction is a function of the concentration of chemical present, the affinity of the receptor to the chemical, and how many receptors it binds to and activates. This relationship between concentration and biological effect is illustrated in Fig. 7.9. In general, this classic *dose-response relationship* shows increasing intensity of a biological effect when the concentration increases. The lower the concentration of drug needed to produce an effect, the more *potent* it is. Each receptor chemical combination has a *response slope* and *maximal effect. This relationship is crucial to understand there must be sufficient chemical concentration present at the site of the specific receptor for an effect to be initiated.* Once all receptors are fully occupied, a maximal effect occurs, where more compound will not produce a greater response. The level of this maximal response is related to efficacy for a drug or toxicity to a toxin. In some cases, this maximal response is insufficient to illicit a systemic effect, for example, when feedback from another system dampens it.

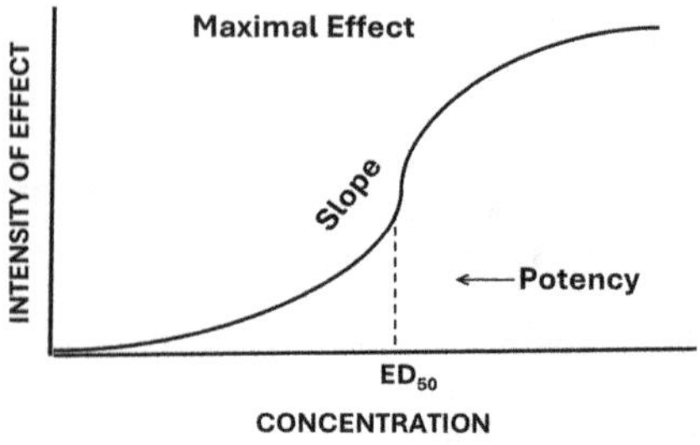

Fig. 7.9 Dose response relationship showing median effective dose (ED_{50})

This dose-response relationship is characterized by the point half-way between zero and maximum effect, the *Median Effective Dose* (ED_{50}). Mathematically it is the inflection point of this sigmoidal function and of all other metrics used to assess potency, it is the most stable. The D-R relationship can also be characterized with the effect axis being the per-cent of a population of individuals showing a specific response with ED_{50} now being the point where 50% of the exposed population shows an effect. This is referred to as a *quantal D-R* relationship. We will revisit this curve, which is central to pharmacology, when we introduce its antithesis, the median lethal dose in the field of toxicology.

There are numerous nuances to this interaction. Some chemicals are antagonists that bind to parts of the receptor with high affinity but do not activate it, such as naloxone blocking the opioid receptor and reversing the action of an opioid drug. Many receptor families have subgroups of receptors with different affinities and levels of effect. Figure 7.10 illustrates this with three different dose-response curves that bind with the same chemical though with different affinities and maximal effect. These curves nicely illustrate some concepts related to defining a zero biological effect. Curve A is the most sensitive receptor, yet has the same maximal effect as Curve C. At a very low concentration (O_2), A will not be activated. As concentration increases, effect increases reaching the maximal effect for that receptor. Compound A is more potent than C. C will only be activated at a higher concentration (0_3). If receptor C is the only one present, it will not be activated at this concentration that would have activated A and thus no effect, a recorded zero response would be observed at that concentration for curve C. At higher concentrations, a maximal effect will ultimately be observed.

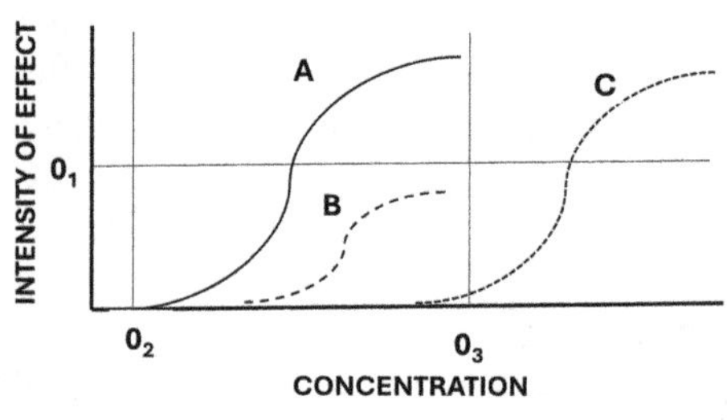

Fig. 7.10 Comparison of different dose response relationships

If both receptors were present, both would be activated and could have an additive effect, for example if the effect was to increase the release of a hormone or some gene product into the blood. This relationship between A and C also can illustrate what occurs after normal drug exposure where A represents the intended therapeutic effect at lower concentrations but when higher levels of drugs are encountered, C activates to produce a toxicologic effect. The difference between the concentrations affecting A and C is referred to as the drug's *margin of safety*. C may also be encountered when drugs are overdosed, and a different response is then activated. In this case, C is the toxicologic response.

Curve B would be a scenario activated at a relatively low concentration but producing a lower maximal effect. If this effect were below that which was detectable or capable of producing a systemic response, shown in Fig. 7.10 as the horizontal effect line (0_1), there would be a zero biological effect at these lower concentrations. Some drugs are only partial agonists and will not induce a maximal response as a full agonist would. Curve B would represent this.

Receptors may become desensitized, that is receptor A essentially behaves as receptor B. In other cases, more receptors are produced when there is chronic exposure amplifying the effect seen. There is also great variability in all aspects of these interactions. The number of combinations and permutations of chemical to receptor interactions are too numerous to document here.

The goal of this discussion is not to present a course on pharmacology or pharmacodynamics. With respect to our discussion, let's return to summarize some examples of *biological nothing* and *zero* at the receptor level.

- If a high concentration of drug is present for a specific receptor such as B, yet no systemic response is seen, then nothing is observed in response to this chemical for this receptor. There is *zero biological effect* (0_1).
- If a true *analytical zero concentration* is present (0_2), there is no drug and there will be no effect.
- If a low concentration is present but below that needed to exert an effect, such as in C, there is *zero biological effect* for C (0_3), even if some low level of binding to the receptor occurred. In this case, both A and B would be activated but these responses may have nothing to do with the effect being studied or measured.

These phenomena also illustrate that the scales used in such dose response relationships are all very subjective and dependent upon drug's potency and the sensitivity of both the analytical instrument that defines the concentration axis as well as the sensitivity of the instrument detecting or observing the biological effect. Some endpoints are reported as a specific fraction of a population, say ED_{90} where 90% of a population would show the effect. There are no absolute unchanging scales and they do not apply across all contexts. For example, a drug may cause toxicity to the heart (cardiotoxicity) that is not detected using a stethoscope but is detected using an electrocardiogram or by measuring a biomarker in the blood.

Up to this point, we have been discussing a single receptor type with exposure of sufficient available drug at the site of the receptor. Such studies often determine the specific potency between a molecule and its receptor. This can be the response measured using *in vitro* cell culture experiments. When we climb the ladder of biological organization to tissues, organs, and organisms; we are often not looking at isolated receptors in a bathing medium, such as would occur in an *in vitro* cell culture experiment, but rather many receptors in different cells often located throughout the tissue or body behind membrane barriers. In cell culture situations, the D-R curve may reflect the actual concentration of a chemical that gets to the receptor as well as the number and sensitivity of receptors present. A downside of cell culture experiments though is that some chemicals may bind to media components which would not occur *in vivo*. Studying intact animals complicates the picture because of distribution to the receptor site discussed earlier as well as systemic responses to the biological effect produced, often dampening it. These will be dealt with in the chapter on toxicology, when specific systems and biological effects are presented. However, when we discuss more complex organismic effects, the fundamental chemical interaction occurs at the level of a single receptor and the sigmoidal dose-response relationship often continues to hold.

Miscellaneous Interactions

Before we leave this discussion, there are a few chemical biological interactions that are different than those discussed above. Immune detection involves high affinity antibodies that can detect extremely low concentrations of chemicals that the body had been previously exposed to. A similar situation occurs with allergic reactions as anyone who has experienced poison ivy can surely relate to. Immune and allergy sensory cells are located on the surface of the skin epithelium specifically to detect such insults. There is chemical binding to a receptor, but not in the graduated nature of the more typical dose response relationship illustrated above.

Sensory systems, where the effect is the firing of a nerve impulse, can be even more sensitive. For example, a classic isolated study showed that when using electrophysiological recordings with silkworm moth antennae, single pheromone molecules could be detected (Sandler et al., 2000). This is not a typical biological or toxicological effect but rather sensory detection. Due to nonlinear effects and cascade phenomenon, a single molecule can be enough for a biological effect. Many *in vitro* systems are optimized for extreme sensitivity. When one moves *in vivo,* the situation changes as all the factors we have discussed may interfere with the molecule ever reaching its receptor.

There are other medical practices in existence that would seem to go against the paradigm presented above. One is *homeopathy,* which is a two-century old alternative medicine practice that uses substances extremely diluted with water or alcohol to purportedly stimulate the body's healing response. The efficacy of this is widely discounted. The next chapter will introduce the concept of *hormesis,* which is a documented "U" shaped dose response where low concentrations of chemicals produce a positive effect before the classic dose response system is invoked. However, this effect seen with multiple chemicals that involves initiating a biological effect that counteracts the toxicological effect at low concentrations, but then is incapable of neutralizing the toxicologic effect produced at higher concentrations.

Summary

This brief overview introduces the biological players involved that are needed to assess the risk of exposure to low concentrations of chemicals in the body. This assessment has a major impact as to what we define as a zero concentration for biological effects. *These details of organismic biology are important because for most compounds, exposure to humans at very low concentrations approaching LOD or LOQ often has no relevance to predicting a significant biological effect because the concentrations needed to activate biologically relevant responses are at much higher concentrations.* This is solely a function of the sensitivity of the analytical technique and biological approach to assess an effect. In other cases, identical receptors are present, but some reside behind protective biological impediments, such as the blood brain barrier that protects the central nervous systems. This complexity generates emergent behavior in the organism. *The nature of the organism's emergent response to a chemical or biological insult, is dependent not only on the normal structural and physiological responses of each cell but also on concomitant factors such as age, presence of disease and nature of the environment.* In many cases, a drug that is harmless to a normal healthy person can be dangerous to a diseased individual whose pathways have been altered from the underlying disease (*e.g.*, hypertensio, diabetes, *etc.*). The risk is very dependent upon the specific physical properties of the chemical that governs its absorption, transport, and distribution as well as its molecular properties related to toxicity and the physiological system being affected. There is no general rule that governs all chemicals.

In many cases, there is simply not enough chemical to be absorbed into the body. In others, they are diluted in the blood and then easily excreted by the kidney. This review is out of necessity simplified and brief but hopefully conveys the complexity of assessing whether exposure to extremely low concentrations of chemicals only detected by extremely sensitive modern analytical methods does NOT always indicate the potential for adverse effects.

Selected Readings

Alberts B, Hopkins K, Johnson A, Morgan D, Roberts K, Walter P, and Heald R. *Essential Cell Biology, 6th Ed.*, New York, Norton, 2023.

Chiou CT, and Block JH. Parameters Affecting the Partition Coefficients of Organic Compounds in Solvent-Water and Lipid-Water Systems. Washington, USGC.gov, 1986.

Haldane T. *Biochemical Pharmacology,* New York, Syrawood, 2016.

Nelson DL. Lehninger Principles of Biochemistry 8th Ed. New York, WH Freeman, 2021.

Riviere JE. *Comparative Pharmacokinetics: Principles, Techniques and Applications,2nd Ed,* Ames, IA, Wiley-Blackwell, 2011.

Riviere JE, and Papich M (eds): *Veterinary Pharmacology and Therapeutics,* 10th Edition, Ames, IA, Wiley-Blackwell, 2018

Sandler BH, Nikonova L, Leal WS, and Clardy J. Sexual attraction in the silkworm moth: structure of the pheromone-binding-protein–bombykol complex. *Chemistry & Biology*. 7:143-51, 2000.

Schrodinger E. *What Is Life? The Physical Aspect of the Living Cell.* Cambridge, Cambridge University Press, 1944.

Urry L, Cain M, Wasserman S, Minorsky PV, and Orr RB. *Campbell Biology, 12th Ed.,* London, Pearson, 2021.

Chapter 8
Toxicology: Dose Makes the Difference

> What is there that is not a poison? All things are poison and nothing (is) without poison. Solely the dose determines that a thing is not a poison. (Paracelsus—(1493–1541))

Up to this point, we have defined what chemicals are, how to detect them, and discussed how they enter the body as well as their fate once absorbed. We presented the concept of the dose-response relationship discussing how the concentration of a chemical or drug is related to the extent and strength of its binding to a receptor expressed as its D-R relationship (Fig. 7.9). At the molecular and cellular level, this response is relatively well-defined and easily measured. This chapter discusses the nature of the adverse responses produced at the level of the organism and population. For example, in the case of a therapeutic cardiac drug, the adverse response may only be an amplification of the desired effect (*e.g.*, further increased heart rate in a cardiac pacemaker cell) or due to activation of a new receptor at the higher dose (*e.g.*, vasoconstriction due to activation of a receptor in an arterial muscle cell causing constriction). How serious must an adverse or toxicological effect be in order to cause concern? A serious effect such as hair loss is tolerated for someone undergoing life-saving chemotherapy but is unacceptable for an over-the-counter headache remedy. We will also discuss how the *magnitude* of the administered dose or exposed chemical concentration, as well as the *duration* of exposure are extremely important.

Humans and all animals evolved in an environment bathed in all sorts of chemicals. They have been constantly exposed to chemical insults. Some compounds were discovered as beneficial and even essential, serving as nutrients and vitamins. Vitamins may be essential elements of a diet at low doses but can be severely toxic at excessive doses (*e.g.*, Vitamin A, D, E, and K). Certain herbs were found to have medicinal properties. However poisonous chemicals were present and identified, especially in plants and mushrooms or as venom produced by poisonous insects and snakes. Animals that survived exposure to these types of chemicals may produce offspring who evolved protective mechanisms to either prevent their absorption or inactivate them through metabolic processes. This is the key driver of Darwinian evolution and natural selection.

Figure Use Attributions
Figs. 8.1, 8.2, 8.3, and 8.4: Wikimedia – No restrictions
All other figures are either self-drawn, photographed or from author's own publications (attributed)

J. E. Riviere, *Zero – Much to Do About Nothing?*,
https://doi.org/10.1007/978-3-031-82998-7_8

The National Institute of Environmental Health Sciences (NIEHS), the US agency where toxicology is its primary mandate, defines toxicology as such:

> The field of science that helps us understand the harmful effects that chemicals, substances, or situations, can have on people, animals, and the environment. Toxicology has been referred to as the "Science of Safety" because as a field it has evolved from a science focused on studying poisons and adverse effects of chemical exposures, to a science devoted to studying drug and chemical safety.

Historical Context

This author contends that this historical transition from toxicology being the science and art of poisons to one of safety is the source of the semantic confusion over interpreting terms ranging from *poisonous, hazardous, having zero effect,* to *nothing happens*. Poison has a persistent negative connotation as clearly evidenced by the images associated with poisonous substances (Fig. 8.1).

Fig. 8.1 Images invoking poisons: The skull and cross bones, the feared scorpion or spider and the suspicious mushroom

In contrast, safety relative to chemical or drug exposure is the lack of an adverse effect, while toxicity or poisoning is when an adverse effect is triggered. These are the two sides to a chemical exposure continuum where concentrations below a lower threshold are considered safe and those above a higher limit are considered toxic. This dichotomous approach of searching for *no effect* versus detection of a *specific effect* defines the terms that are used for *safe limits of exposure* or *maximum tolerable dose.* For a therapeutic drug, this would be what was presented as a *margin of safety*. The use of so many terms, for related but not identical safety or toxicological endpoints, spreads chaos when interpreting values out of the context of how the terms were calculated and defined.

It is obvious that ancients and even our prehistoric ancestors knew of the existence of poisons, both in the form of plants, mushrooms, insects, and snakes to avoid; as well as substances useful to assassinate one's enemies. In India, the *Ayurveda* traced back to 5000 BCE lists plant-based treatments for medicinal purposes. The earliest compilation of medicinal substances was the ancient Chinese *Pen Tsao* attributed to Emperor Shennong in 2700 BCE, listing some 365 medicinal substances. Not surprisingly, he is attributed to inventing tea! The Egyptian *Ebers Papyrus* from 1150 BCE contained 800 prescriptions for salves, plasters, pills, suppositories, and other dosage forms to treat specific ailments.

The earliest true *Materia Medica*, a compilation of therapeutic substances, was compiled in 77 CE by Aristotle's student *Dioscorides,* who served as a surgeon in Emperor Nero's Roman Legion travelling through the Mediterranean. Nero himself conducted experiments on slaves to ensure poisons used by his assassins targeting his enemies were effective (Fig. 8.2).

Fig. 8.2 Nero having Locusta testing the poison prepared for Britannicus. On the ground is the slave who had to try the poison

This earlier work formed the basis for the compilation of Galen that became the authoritative Material Medica for 1400 years. The Iranian physician Avicenna also published books containing medicinal substances around 1000 CE, integrating the previous works of the Greek and Roman physicians. The first printed, and more easily accessible, translated versions of Dioscorides appeared in Latin in 1478 and in Greek in 1499. The dawn of the sixteenth century saw a proliferation of such compilations of Materia Medica throughout Europe. In the modern era, the Materia Medicas have been replaced by scientifically more rigorous *Pharmacopeias* issued by national authorities. It is interesting to note that the first edition of the U.S. Pharmacopeia published in 1820 is referred by some as volume zero!

Paracelsus, the alchemist we met earlier, was known for incorporating metals into contemporary Material Medicas. However, he is remembered today for being responsible with the quotation leading off this chapter that *the dose makes the poison,* and for the need to conduct experiments to determine a biological effect. He experimented with various potions and heavy metals, including mercury, lead, arsenic, and antimony to cure ailments of the day. He was recognized for using drugs for specific and directed purposes, for example, mercury as the drug of choice for syphilis, a treatment that persisted for three centuries. He also introduced the clinical use of tinctures (alcohol extracts) from various plants to treat many diseases. This included the opiate laudanum (opium) as a pain killer, similar to paregoric used in the past 50 years to rub on infant gums while teething. Some remedies worked and others didn't or even maimed or killed his patients, giving such practitioners today the moniker *poisoners. However, Paracelsus' observations clearly demonstrated the dose-response relationship that formed the basis of the field of toxicology from that point onward.*

During the ensuing centuries, the rapid development of chemistry, as outlined in Chapter Six, greatly increased our understanding of the properties of metals and chemicals as well as our ability to detect them. This work led to the development of modern pharmacology in the late nineteenth and early twentieth centuries. Pioneers such as Paul Ehrlich in 1910, developed arsenic containing antibiotic arsphenamine (salvarsan) as the magic bullet for treatment of syphilis as well as for other parasitic diseases including trypanosomiasis. He also articulated the safety margin or therapeutic index. This fascinating story is a topic for a different book! (Fig. 8.3)

Fig. 8.3 Paul Ehrlich

During this period, occupational hazards encountered by metalworkers (heavy metals like mercury and lead) and chimney sweepers (polyaromatic hydrocarbons—PAHs in soot leading to scrotal cancers) led to the development of the field of industrial toxicology and emergence of occupational regulations in England, Germany, and the United States around the turn of the century. During this period, chemical warfare agents including phosgene and mustard gas, widely used in the First World War in Europe, as well as drugs, such as aspirin and steroids, and agricultural chemicals including pesticides like DDT, were synthesized by organic chemists.

At the turn of the twentieth century, the Nobel prize winning German chemist Fritz Haber (1868–1934) (Fig. 8.4) formulated *Haber's Law, which states that the incidence and/or severity of a toxic effect depends on the total exposure, that is exposure concentration (c) rate times the duration time (t) of exposure (c × t).* This relationship characterized the acute toxicity of gas warfare agents but has formed the basis of setting chemical exposure guidelines for toxic substances in the second half of the twentieth century. Although referred to as a law or rule, it is not precise in all applications, but conveys the valid concept that *dose is a function of both concentration and length of exposure*. It is used in establishing reference doses (acceptable daily intakes) for long-term exposures when only the results of short-term studies are available, though an uncertainty (safety) factor discussed later in this chapter is added to ensure safety. There are several concerns with repeated or chronic chemical exposure; including accumulation of compounds in the body or select tissues, induction or inhibition of enzymatic metabolism, and initiation of both positive and negative feedback processes. As will be seen, these factors are now accounted for in modern computational risk assessment algorithms. Similarly, in pharmacology, a *dosage regimen is defined as the dose of drug as well as its frequency of administration* (*e.g.*, one tablet once a day, twice a day, *etc.*). *The reader should always consider a dose as being both mass of drug administered and frequency of administration.*

Fig. 8.4 Fritz Haber

Chemical Risk Assessment

Before we dive into the details of the practice of modern toxicology, it is a good point to discuss how the adverse risks to chemical exposure are evaluated. We discussed the general concept that the combination of dose and duration of a chemical exposure is related to its potential toxicological effect. The second major determinant is the nature of and potency of the chemical's inherent toxicity, that is "what does it do?"

An agent's *mode of action (MOA) is the group of biological and physiological effects that define the nature of the adverse response after compound exposure.* This could range from a chemical's binding to a specific receptor that induces an adverse effect, a change that induces a physiological feedback response, or in some cases a direct alteration in gene expression. Some MOAs are very specific to certain cells or tissues, while others nonspecifically damage any biological tissue. Examples would include a potent neurotoxin targeted to a specific nerve cell receptor versus a caustic chemical, such as a strong acid, that injures everything it encounters.

The one missing piece of this puzzle is whether any exposure to the chemical actually occurs. The process of assessing and quantitating the potential for adverse events after chemical exposure is termed *risk assessment*. Simply stated:

$$\textbf{Risk} = \textbf{Hazard} \times \textbf{Exposure}$$

The first step in the process is *hazard identification*: does the chemical produce any harm? A *Weight of Evidence (WOE)* summary, defining the relationship between the chemical and its associated toxic effects, is written based on examining available animal, human epidemiological, and *in vitro* test and *in silico* model data. If there is a concern, then the MOA is further defined, and a dose-response relationship analysis conducted. In the past, these studies were only performed on laboratory animals as described in the next section. However, today, more sophisticated *in vitro* and *in silico* approaches, including alternative animal models, have been developed. Toxicologists have become skilled in identifying laboratory animal species and strains which mimic specific toxicological responses in humans. In the National Research Council's paradigm termed *Toxicity Testing in the 21st Century*, toxicity pathways are defined using human-relevant *in vitro* systems coupled with mathematical *in silico* models that describe both chemical disposition (ADME processes) and activity, topics that were introduced in the previous chapter.

At this stage, one should have a reasonable grasp of the potential hazard of the chemical. The next required step is to document exposure. Can humans be exposed to this hazardous chemical? What is the likely human intake? What are the routes of and scenarios for significant exposure? An analysis is conducted quantitating what the actual exposure is by all relevant routes. Examples include daily food intake of products containing this chemical, occupational and household exposure, consumption of polluted water, exposure of a compound topically applied on skin, or breathing of toxic gases in the air. Exposure to a toxic pesticide is very different for the casual gardener compared to the agricultural worker spraying hundred-acre fields daily; same chemical, same potential toxicity, but vastly different exposures with cumulative effects possible over periods of months to years. This is the same issue often seen with occupational exposures.

Once the exposure scenario is precisely defined in terms of dose and duration, this is compared to the appropriate dose metric that links dose to various levels of toxicity. The construction of such dose metrics will be discussed fully in the sections below and range from no effect to lethality. The specific legally allowable thresholds are very dependent upon the country promulgating them and the specific regulatory agency involved. The metrics should be determined using the same route and matrix as the potential exposure to account for differences in bioavailability. *Uncertainty factors* are added to the risk assessment to lower the acceptable dose to account for variability. These uncertainty factors are very dependent upon the type of chemical studied, the amount of data available to define the D-R relationship, the nature of the D-R relationship, the severity of the toxicological response and the availability of human epidemiological data reporting the effects of exposure, and the adverse effect observed.

If the exposure is greater than the relevant dose metric, then there is a potential for adverse effects to occur. If the exposure is well below the metric, then adverse effects from exposure to the chemical are unlikely (Fig. 8.5).

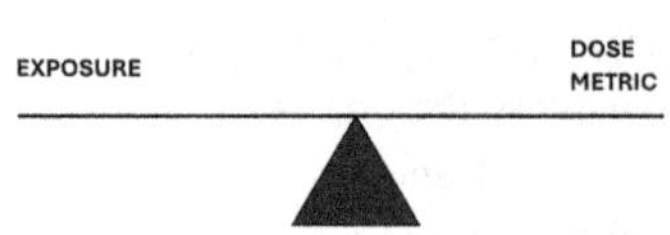

Fig. 8.5 The need to balance estimated exposure to the dose that causes a defined adverse effect

As this figure illustrates, *risk assessment is weighing exposure to a chemical against what adverse effects that level of exposure would cause to a human exposed to it.* The bulk of the science of toxicology entails defining all the variables that allow an accurate characterization, with an acceptable risk, to be made. In the US, agencies such as the Agency for Toxic Substances and Disease Registry (ATSDR), the Environmental Protection Agency (EPA), the Food and Drug administration (FDA), and the Occupational Safety and Health Administration (OSHA), conduct these risk assessments.

The Goal Posts of Quantitative Toxicology: LD_{50} and NOAEL

Now that we have a basic idea of what the background and goals of toxicology are, it is time to focus more on how toxicological effects are quantitated. There are two parameters that will serve as our goalposts, one demarking maximum effect or lethality and the other demarking low to no effect.

Median Lethal Dose—LD_{50} The primary dose metric of the *maximum potency* of a toxic substance for a large part of the twentieth century was related to mortality as the endpoint of concern. *This metric of chemical lethality was defined as the single dose of chemical that resulted in the death of 50% of the animals administered that dose, a metric termed the median lethal dose or LD_{50},* first coined by JW Trevan in 1927. The LD_{50} was determined by administering escalating doses of a test chemical to groups of laboratory mice and rats. The shape of this dose-lethality curve was generally the now familiar sigmoidal plot we encountered with the dose-response relationships in the biology chapter (Fig. 7.9), with ED_{50} now becoming LD_{50}. In this case, the effect axis is now the mortality axis showing the per cent of the population dying after acute chemical exposure. The emphasis on the use of such acute rodent toxicity testing has changed dramatically over the past decades. Precise estimates of an LD_{50} are often no longer pursued because limited data showing no deaths at large grams/kilogram doses *via* the route of expected human exposure are sufficient for assessing the safety of many industrial chemicals. This can be accomplished by dosing small groups (6–10) of animals and increasing doses by various strategies until death is observed or a selected pre-defined "large" gram/kilogram dose is achieved. There is no need to completely map out the D-R lethality curve because exposure greater than the "large" dose relevant for a specific chemical is impossible. This approach is far more humane than earlier practices.

Numerous factors, including age, gender, strain, breed, nutritional status, housing, dosing vehicle, route of administration and diet can affect these estimates. Animal strain can be especially important as values may change greatly depending on the genetic strain used. The author encountered this years ago when studying the renal toxicity of the antimicrobial gentamicin in Sprague Dawley *versus* Fischer 344 rats, with the latter being an order of magnitude more sensitive to kidney damage after administering the same dose of drug. Selectively bred genetically defined strains of mice show widely variable LD_{50} values for numerous compounds, making historical comparisons difficult and underlying the strong genetic bases of sensitivity to poisons. This fact, coupled with more sophisticated *in vitro* and *in silico* approaches available today, as well as with our modern attitudes toward animal welfare, have reduced acute animal toxicology trials significantly. When they are conducted, they use a lower number of animals, select animals with a response known to mimic that seen in humans, and escalate doses until lethality is observed.

Table 8.1 lists the oral LD_{50} for a range of chemicals in laboratory rats. The tabulated LD_{50} is generally listed as a mass of the drug per unit of body weight. From an acute toxicological perspective, they represent the potency of different toxins. There are a few surprises in this table. Classic poisons, including cyanide and strychnine, are lethal at single digit mg/kg doses. Various poisonous snake venoms' lethality generally ranges from approximately 0.1–10 mg/kg, though the Australian small-scaled snake (*Oxyuranis microlepidotus*) has an LD_{50} of only 0.01 mg/kg. Chemical warfare nerve agents and some mycotoxins produced by fungi (poisons such as aflatoxin and tetrodotoxin) fall into this range and are lethal at sub mg/kg levels. The most potent toxins are the contaminant dioxin and botulinum toxin produced by the anaerobic bacterium *Clostridium botulinum.* It's curious that despite the extreme potency of botulinum, it is widely used in pharmaceutical formulations for cosmetic purposes (Botox®)! One of the reasons for the magnitude of the current illicit fentanyl crisis can be appreciated when comparing potency values for morphine *versus* cocaine *versus* fentanyl.

Table 8.1 LD_{50} of common chemicals in laboratory rats

Toxic chemical	LD_{50} (mg/kg) (PPM)
Water	90,000
Ethanol	7000
Salt	3000
Morphine	900
Caffeine	400
Phenobarbital	150
Cocaine	100
Cyanide	5
Strychnine	5
Fentanyl	3
Polychlorinated biphenyls-PCB	0.1–1
Aflatoxin B1	0.5
VX nerve gas	0.1
Tetrodotoxin	0.1
Dioxin (TCDD)	0.001
Botulinum toxin	0.00001

As discussed in the NIEHS definition of toxicology, this field evolved from defining toxicity in terms of a severe endpoint such as death, to less obvious endpoints of organ dysfunction and biochemical markers of adverse effects such as elevated cholesterol or high blood pressure. LD_{50} or some acute measure of "large" dose toxicity are still used in hazardous material labelling and on the globally harmonized Safety Data Sheets (SDS) for industrial chemicals, such as cleaning solvents and agricultural chemicals like pesticides.

No Observed Adverse Effect Level—NOAEL As the science of chemical and drug toxicology continued to develop through the second half of the twentieth century, large-scale chemical and drug exposure trials in laboratory animals focused on sub-lethal endpoints and the development of metrics to define the dose that produced an absence of specific events in place of lethality, that is safety *versus* toxicity. The primary parameter encompassing this concept is the *No Observed Adverse Effect Level—NOAEL or NOEL. The NOAEL is the largest dose of chemical experimentally found to cause no detectable adverse effects in laboratory animals.*

Depending on the purpose of the study, toxicologists usually define a working definition of injury as *any alteration that diminishes essential biological functions whether or not cell death ensues.* Cell death is not always undesirable *as programmed cell death termed apoptosis* is the normal process the body uses to replace older or damaged cells. Recall that the human body loses some 60 billion cells every day, often due to apoptotic pruning or cell death from scenarios such as sunburns. Many tissues have excellent regenerative capability. Additionally, many biological responses may be adaptive changes to combat chemical exposure. This response will be further discussed when *hormesis* is presented. *The biological significance of any observed change is a value judgement made by an individual investigator or consensus panel of regulatory scientists.*

The endpoints observed are comprehensive and change with developing science, but include mortality, changes in body weight, reduced food consumption, behavior (*e.g.*, activity level, sleep *etc.*), hematology, immunological tests, blood and urine clinical chemistry markers of organ function, as well as pathological and microscopic (histological) changes seen at necroscopy (an autopsy for animals).

In all these studies, a broad range of doses are included as well as control animals administered the vehicle without study chemical (*zero dose*). The studies are conducted under *Good Laboratory Practices (GLP)* regulations to ensure consistency, quality, and accountability of the results. There are specific study designs to define acute (see LD_{50} discussions above), subacute (14 days), subchronic (10% of animal lifespan), and chronic (animal lifespan) durations depending on the relevant regulatory agency, thereby consistent with Haber's Law. Chronic studies may also include endpoints of carcinogenicity. Some studies are multigenerational, which assesses reproductive toxicity in the parents and mutagenic, genotoxic, or teratogenic effects in their offspring.

Subchronic studies are generally used to determine the NOAEL, however, for compounds where the effects of long-term exposure are desired, longer term and often multi-generational studies may be conducted. For the typical subchronic study, dosing is either daily or five times per week over the duration of 10% of an animal's lifespan, for example 90 days in rats. A control (no test chemical, just vehicle) and at least three dose groups are used with the highest dose resulting in less than 10% mortality. The subacute toxicology study is used to estimate this upper dose. Enough animals per dose group (two species, including both males and females) are employed to ensure statistically valid conclusions for both sexes. The route of administration matches the exposure route being modeled in humans. In such experimental studies, the lowest dose where a biological effect is detected is termed the *lowest observed adverse effect level* (*LOAEL*). The actual experimental design of statistically valid toxicology studies is well beyond the scope of this book, with the primary point being that the parameters reported are considered valid.

Perspective on LD_{50}, NOAEL, Units and Biological Zero One can view the progression of toxicological changes seen with increasing chemical dose as:

HOMEOSTASIS → REVERSIBLE DAMAGE → IRREVERSIBLE DAMAGE

This line could be annotated by adding our two "goalposts" metrics that occupy the boundary "book ends" to this toxicology dose line:

NOAEL → HOMEOSTASIS → REVERSIBLE DAMAGE → IRREVERSIBLE DAMAGE → ***LD_{50}***

The divider between homeostasis and slight reversible damage is referred to as the *threshold of safety* for a chemical. For human drug approvals, clinical trials are conducted by FDA to precisely define safe dosing levels (Phase One clinical trials), which are performed before larger scale efficacy trials (Phase Two and Three) are conducted to identify both the label dosage regimen and identify sensitive populations. When discussing complex biological responses, individual factors such as genetics, age, disease, and previous or concomitant chemical or drug, may shift this line. Some individuals in a population are sensitive to specific toxins while others are resistant, a finding holding for most biological effects. *Variability is the hallmark of biology.* Post-approval monitoring (Phase Four) assures safety in the larger general population now using this drug.

There is one other point to make about toxicological potency metrics, such as LD_{50} and NOAEL; they are expressed in units of mass per body weight, generally mg/kg (ppm) or μg/kg (ppb). Another perspective on assessing potency is to instead express them based on number of molecules exposed, that is the molarity of the chemical or drug exposure, generally mmol/kg (see chemistry chapter for review). Table 8.2 compares the potency of two hypothetical chemicals differing in molecular weight 10–100-fold. Note the large differences in potency when expressed in this manner. Molar doses are often used in mechanistic studies where equal ratios of moles of substances need to be compared.

Table 8.2 Comparison of potency of two molecules based on concentration (Mass of chemical) versus molar exposure (Number of chemical molecules)

Molecular weight		Toxic dose A		Toxic dose B		More toxic
A	B	mmol/kg	mg/kg	mmol/kg	mg/kg	Chemical
1000	100	0.01	10	1.0	100	A
1000	1000	0.01	10	0.1	100	A
100	1000	0.1	10	0.1	100	A = B
100	10,000	0.1	10	0.01	100	B

This is a good time to point out the differences in drug approval *versus* risk assessment of environmental, drinking water and food contaminants where clinical trials cannot be conducted. *Drugs are generally approved only after their safety at specific dosage regimens has been demonstrated in the four-phase clinical trial paradigm.* For some drugs, such as those used to treat rare diseases, they undergo shorter trials because effective therapies may not be available. In contrast, for contaminants, human clinical trial data are not available. This requires conducting laboratory animal studies, *in vitro* studies, and computational approaches to predict what the effect in humans might be. A great deal of research has determined the appropriate animal model in addition to rats and mice to model specific toxicological endpoints (e.g. guinea pigs, rabbits, etc.). Human epidemiological analyses are often employed here, a topic that is further developed in the next chapter on carcinogens. Appropriate uncertainty / safety factors are then employed. *Compared to the world of carefully controlled human clinical drug approval trials, uncertainty prevails in assessing risk of many contaminants.*

To complete the discussion of biologically relevant metrics for chemical and drug exposure, bioactive drug and venom concentrations are expressed in units of activity, which correlate more with molarity than mass. Such units, often internationally standardized as *International Units* (IU), express the chemical quantity in terms of *units of biological activity* required to exert a specific effect. Common substances quantitated in IUs are drugs, such as the antibiotics penicillin and gentamicin, vitamins, and numerous snake and insect venoms. Historically, antibiotic IUs were used to measure concentrations of bioactive chemicals that were not available in pure chemical form; and before synthetic and analytical chemists were able to provide better quantitated products. When these lethal poisons with very narrow margins of safety were employed, knowing precise doses in terms of activity were essential before analytical chemistry methods to quantitate these substances became routine. To calculate actual concentrations, IUs may be converted to units of mass. The use of such units allow compounds, such as venoms, to be incorporated into drugs and are still widely used for snake and insect venom potency.

Essential Nutrients and Hormesis

Some compounds, including many nutrients and vitamins essential to normal physiological functions, are required by the body and thus can show adverse effects both at very low (deficiency) and very high (toxicologic) doses. Instead of the classic sigmoidal D-R relationship that shows no to very low effect at low doses (Fig. 7.9), these compounds may have a U-, inverted U-, or so-called J- shaped D-R curve as seen in Fig. 8.6. The solid line (1) is typical of a vitamin or essential nutrient where deficiency at low doses creates adverse effects, due to vitamin deficiency including death, while at very high doses, toxicity may result. In contrast, the dashed line (2) shows an effect at low doses, that is lost at middle doses and returns at high doses (hormesis response), while the dotted line (3) shows mid-doses having the greatest effect (e.g. selenium, some endocrine disruptors).

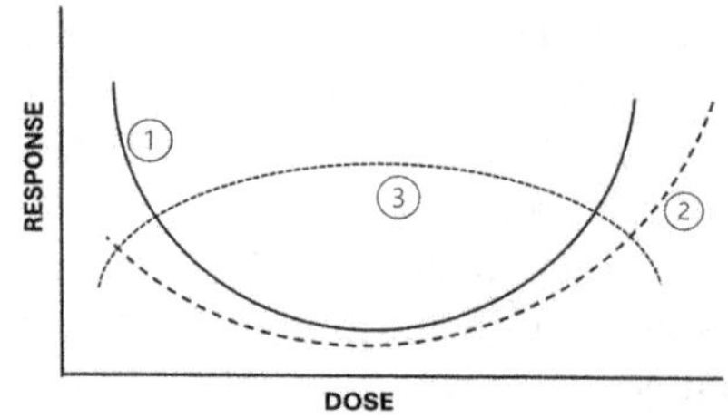

Fig. 8.6 U-shaped D-R curves. Curve 1 would be typical of a vitamin, curve 2 illustrates a low dose reduction and high dose enhancement hormesis while curve 3 shows a low dose stimulatory and high dose inhibitory hormetic response

Hormesis is formally defined as a response curve where a beneficial stimulation occurs at low doses that at higher doses is overwhelmed and produces an adverse effect. These effects are complex and very compound dependent. They result from the chemical agent inducing different processes at different doses, with the resulting D-R curve reflecting a sum of all processes. Some of these effects could be inhibitory to the chemical's adverse effect while others could stimulate the process. Low dose exposure to some chemicals may induce protective metabolic responses that are beneficial overall to the organism. However, when these defensive processes are overwhelmed, adverse effects then occur. This phenomenon, when present, creates difficulty extrapolating low dose effects from data collected at higher doses, or when integrating data from isolated *in vitro* systems where such hormetic effects may not be seen. This problem will be discussed further in the chapter on cancer.

Additional Factors in Toxicological Assessments

There are many confounding factors in defining safe levels of chemical exposure to humans. These include:

- Age related sensitivity in the very young and old
- Genetic differences in tissue response
- Differences in ability to metabolize compounds (both inactivate and activate)
- Pre-existing disease that might exacerbate a toxicological response
- Lifestyle factors such as alcohol consumption, smoking, overeating, illicit drug use, etc.
- Exposure to a chemical in a mixture
- Exposure to allergens in sensitive individuals (*e.g.,* anaphylaxis in peanut allergies)
- Abnormal individual consumption (e.g. eating only one type of food, excessive water consumption, *etc.,*)

Very young and old individuals often have deficiencies in excreting or metabolizing drugs and chemicals in their livers or kidneys due to not being developed in the young or being senescent in the old. The American Geriatric Society has published guidelines for drugs that are potentially harmful to older people. The very young are undergoing development in numerous body systems that may be sensitive to some chemical exposures (*e.g.*, lead affecting neurological development). The very old are often taking multiple pharmaceutics and supplements and may have concomitant diseases making further organ dysfunction dangerous. They have accumulated a lifetime of chemical insults that may predispose them to adverse effects upon further chemical exposure. Similar scenarios occur for individuals who have liver dysfunction secondary to alcoholism, or lung dysfunction from chronic smoking. Finally, there are very specific genetic predispositions to certain chemicals in affected individuals that are very difficult to consider when setting population exposure limits. These can be limited to specific chemical classes or may be expressed as deficiencies in specific metabolic pathways that accentuate potential toxicity. An example is many East Asians have an inability to metabolize alcohol due to decreased alcohol dehydrogenase activity.

Chemicals are rarely exposed as single chemicals and often are found in complex chemical mixtures. Examples of such compounds are PCBs and PFAS. Exposure limits are often set to a sum of potential individual chemicals. This presents a challenge to interpret individual chemical concentrations. In other cases, chemicals may occur in a mixture containing compounds that accentuate absorption or metabolism of another component in the mixture. Individuals with allergies must always be vigilant to ensure they are not exposed to specific compounds to which they are sensitized. For drugs and foods, this is often accompanied using warning labels. Recent work by the author has also illustrated how unique adverse drug effects, not normally seen when the drug is administered alone, occur due to multiple drug interactions with other medicines being administered. These are hard to define for drugs even when dosing data are available. For chemical exposures, such data are just not available.

All human exposures are based on population estimates for the average individual, with variation from the norm accounted for using safety factors as described below. However, issues can always arise if an individual consumes an unbalanced diet consisting of either a single food or obtaining food from only one source that happens to be contaminated. Eating a radically unbalanced diet can expose one to higher levels of potential toxins and induce dietary deficiencies that themselves can produce adverse effects. Similarly, some compounds accumulate in the body after prolonged exposure and may only cause toxicity after reaching a certain accumulation. If a compound accumulates in adipose tissue and an individual goes on a diet and loses significant weight, higher concentrations may then be present due to redistribution of the compound from other tissues and produce an adverse effect. A useful analogy is that of a full barrel of water where one more drop causes overflow. The effect of an exposure today is a function of how much compound one has already accumulated. This phenomenon explains why chemicals prone to accumulation are often regulated at very low exposure concentrations. These scenarios, with examples, will be fully discussed in the food safety chapter.

Most toxicological risk assessments are based on animal studies, alternative animal models, human *in vitro* cells, and recently, organ culture experiments. *In vitro* studies (*e.g.*, cell or organ cultures) involve taking specific cells or tissues and maintaining them in the laboratory. Chemicals can be directly applied to the tissues and cellular growth or death, biochemical markers of toxicity, or other responses measured. What is lacking in these systems are the ADME processes that deliver the compound to the cell. It is also difficult to obtain concentration profiles that mimic *in vivo* conditions. They also lack an intact organism's ability to compensate for mild individual cellular dysfunction and lack the normal tissues that protect these cells *in vivo*. *In general, in vitro tests tremendously overestimate the potential for systemic toxicity from low level chemical exposure.*

When available, human epidemiological data often can elucidate whether chemical exposure to a population result in adverse effects. A larger and more representative population results in a better prediction. A weakness of these studies is often the lack of an accurate estimation of actual exposure through measuring intake. It is also essential that other factors that may affect the outcome are properly controlled.

In the last decade significant progress has been made in integrating *in vitro*, *in vivo* animal, and human epidemiological studies into comprehensive computational platforms that better simulate adverse effects, taking all sources of experimental data and historic exposures into account. This is the goal of the previously mentioned *Toxicity Testing in the 21st Century* initiative.

As I am sure the reader appreciates, toxicology is a broad discipline whose goal is to understand the potential for chemicals to cause adverse effects in humans. It is an evolving discipline where new techniques and data are constantly being developed. As seen for most science and technological applications, the legal regulatory framework is based on science and technology that now may be decades outdated.

This is especially relevant in the advances in sophisticated *in vitro* approaches, complex computational models incorporating artificial intelligence, and ever more sensitive and rapid analytical chemistry assays.

Quantitative Toxicological Endpoints

Similar to what we encountered when discussing metrics of lower limits of assays in analytical chemistry (LOD, LOQ), and our "bookend" values for animal toxicology studies (LD_{50}. NOAEL), there are numerous legal thresholds established based on the observation of minimum biological effects used to regulate chemical and drug exposure to humans. The specific endpoints are very dependent upon the regulatory agency involved (*e.g.*, FDA, ATSDR, OSHA *etc.*) and the laws that govern them as each agency has a specific regulatory role in protecting the population. These include:

- Bench Mark Dose (BMD)
- Biological Exposure Indices (BEIs)
- De Minimis Level — < 0.5–1 ppb
- Minimal Toxic Level (MTL)
- Minimum Level of Applicability (MLA)
- Permissible Exposure Limit (PEL)
- Reference Concentration (RfC)
- Reference Dose (RfD)
- Threshold Limit Value (TLV)
- Threshold of Regulatory Policy
- Threshold of Toxicological Concern (TTC) (< 1.5 μg/day intake)

Several other metrics are used for chemical and drug residue consumption in food, a topic to be fully discussed in Chapter Ten on food safety. All these metrics attempt to translate the experimentally obtained NOAEL to define exposure scenarios of concern and thus establish tolerances to limit exposure that are well below the NOAEL. A few of the metrics above will be discussed to provide some context as to different bases for their determination. Finally, the legal regulatory process is slow, and the endpoints are often dated and based on the science that was current when the metrics were established. This is especially pertinent with the rapid advances in analytical chemistry that result in chemical detection being much more sensitive to what was available when many of the metrics were established.

In the past few years, some regulatory agencies have used the concept of the *Benchmark Dose (BMD),* rather than NOAEL, to estimate the allowable exposure limits of substances more accurately, including carcinogens, by combining multiple models and better estimates of uncertainty. The *BMD is a dose level associated with a specified change in response relative to the control group or background response that is deemed an acceptable level of adverse effect, for example a response at 1 or*

10% of the population showing a response. This clearly depends on the severity of the adverse effect being studied. This method involves fitting a mathematical model to all the dose-response data within a study and incorporating available biological information into the resulting estimates of exposure limits. Some agencies use the RfD or RfC to identify the maximum dose or concentration of a compound that can be tolerated by daily exposure in humans. Other agencies continue to use the NOAEL. There is a fundamental difference between these two metrics. *The BMD uses mathematical dose-response models to estimate a response, in contrast to the NOAEL which is solely based on observed experimental laboratory animal data.* For our purposes, they both represent endpoints of lack of biological effect after chemical exposure, our *operational zero*.

Threshold Limit Values (TLVs) are the maximum average concentration of a hazardous material in the air that can be inhaled for five consecutive eight-hour workdays without causing significant health issues for healthy adult workers. The TLVs are developed by the American Conference of Governmental Industrial Hygienists (ACGIH) and are based on studies of human and animal exposure. Occupational exposure thresholds, such as TLV, are based on exposure over a normal work week. In contrast, RfC is the concentration determined to be safe over a lifetime inhalational exposure.

Other limits are set depending on the nature of the exposure events. Levels in drinking water are based on normal daily water consumption values. The amount of water needed to be consumed to reach the NOAEL, or a BMD is calculated and then divided by safety factors due to uncertainties. Most published exposure limits have safety factors built into their calculation and are conservative estimates of determining the no effect levels (See Fig. 8.7).

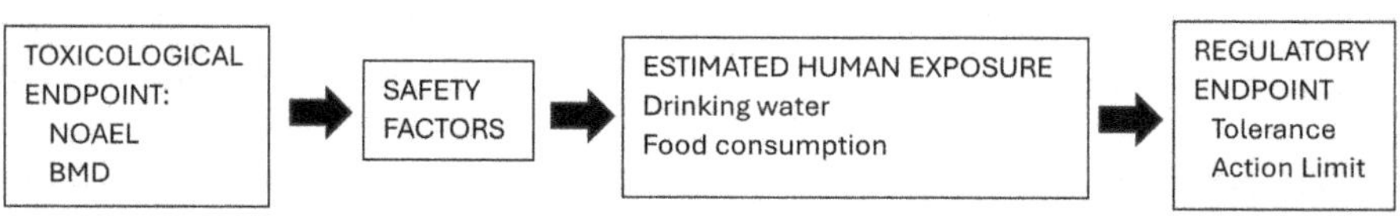

Fig. 8.7 Establishment of allowable thresholds

In many cases where data are limited, regulators have attempted to define if there are minimal exposures of any chemical, for example most organic pesticides or some environmental contaminants, required to produce an adverse effect. These are expressed as the *De Minimis Level, MTL, or TLC. Levels of less than 0.5 ppb in food or water are often considered of no toxicological concern. Expressed as a dosage to humans, the value may be an intake of less than 1.5 μg/day. These are extremely low numbers*. Some workers argue that these low levels do not hold for compounds, such as carcinogens (to be dealt with in the next chapter), some protein allergens, and some metals including lead, which can be dangerous for children undergoing development of their nervous systems. Levels for PFAS (per- and polyfluoroalkyl substances), the so-called "forever chemicals" have been set as 70 ppt for a *combination* of these compounds in drinking water.

The concept behind these minimum levels is based on the belief that a sufficient concentration of a chemical may be required to be absorbed, distributed through the body and cellular transport systems to reach a target receptor. Dilution, diffusion, and metabolism require a sufficiently large exposure mass of chemicals to drive these processes. Even allergic or immunologic responses require a concentration sufficient to illicit a response to be present. The body's defensive responses must also be taken into consideration.

Depending on the nature of the compound being regulated and by whom, exposure limits and tolerances, such as discussed above, are set and used to monitor human exposure to these compounds. All involve the distinct steps of defining a "*safe level*" from toxicological studies and depending on both the uncertainty of the data, factoring in safety factors that further reduce the allowable concentration and then estimating human exposure to set final allowable concentrations. For some putative carcinogens, the safety factor can be as high as 1000×. This process is illustrated in Fig. 8.7.

Conclusion

As one can appreciate, the science of toxicology is a complex endeavor that encompasses multiple disciplines. The focus of this chapter was not to give a course in toxicology, but rather to introduce some key concepts of chemical safety assessment to give the reader a perspective of what a safe concentration of a chemical is and how it was determined. Importantly, it gives us an operational definition of *zero* and *nothing* relative to the absence of adverse biological effects.

The NOAEL is the dose of a chemical that is statistically equivalent to the responses seen in the control group of animals not administered any chemical. The NOAEL represents our "Holy Grail of Zero Biological Effect". One must understand that animals exposed at the NOAEL dose show no observable adverse effects, yet they actually received a low dose of chemical that can often be analytically quantitated! Recall, it is at the higher LOAEL where adverse effects were observed. *Exposure to a drug, that at higher doses can produce adverse effects, may be given at a dose less than the NOAEL with no effect.* This is the scenario that happens when one takes most approved drugs. At the dose given, a beneficial effect will be seen. At higher doses, adverse effects may be observed. Finally, potency of a chemical is not equivalent to risk. It is just one factor that determines the toxic dose. Even a low exposure to a potent chemical may be below its NOAEL.

Exposure does not equate to toxicity

This simple fact is essential to fully appreciate because detection of a chemical in water or food does not mean that its consumption will result in toxicity, especially when a chemical is detected only at its LOQ or LOD. This is especially true today because these safe values are often well below the NOAEL using modern analytical techniques.

Selected Readings

American Geriatrics Society 2015 Beers Criteria Update Expert Panel, Fick DM, Semla TP, Beizer J, Brandt N, Dombrowski R, DuBeau CE, Eisenberg W, Epplin JJ, Flanagan N, and Giovannetti E. American Geriatrics Society 2015 updated beers criteria for potentially inappropriate medication use in older adults. *Journal of the American Geriatrics Society*. 63:2227-46, 2015.

Calabrese EJ, Bachmann KA, Bailer AJ, Bolger PM, Borak J, Cai L, Cedergreen N, Cherian MJ, Chiueh CC, Clarkson TW, Cook RR, Diamond DM, Doolittle DJ, Dorato MA, Duke SO, Feinendegen L, Gardner DE, Hart RW, Hastings KL, Hayes AW, Hoffmann GR, Ives JA, Jaworowski Z, Johnson TE, Jonas WB, Kaminski NE, Keller JG, Klaunig JE, Knudsen TB, Kozumbo WJ, Lettieri T, Liu SZ, Maisseu A, Maynard KI, Masoro EJ, McClellan RO, Mehendale HM, Mothersill C, Newlin DB, Nigg HN, Oehme FW, Phalen RF, Philbert MA, Rattan SIS, Riviere JE, Rodricks J, Sapolsky RM, Scott BR, Seymour C, Sinclair DA, Smith-Sonneborn J, Snow ET, Spear L, Stevenson DE, Thomas Y, Tubiana M, Williams GM, and Mattson MP. Biological stress response terminology: Integrating the concepts of adaptive response and preconditioning stress within a hormetic dose-response framework. *Toxicology Applied Pharmacology.* 222: 122-128, 2007.

Frazier DL, Carver MP, Dix LP, Thompson CA, and Riviere JE. Exaggerated response to gentamicin-induced nephrotoxicity in Sprague Dawley rats: Identification of a highly sensitive outlier population. *Toxicological Pathology.* 14:204-209, 1986.

Gaylor DW. The use of Haber's law in standard setting and risk assessment. *Toxicology*. 149:17-19, 2000.

Hayes AW, and Kobets T. *Hayes' Principles and Methods of Toxicology, 7th Ed.* Boca Raton, CRC, 2023.

Klassen C. *Casarett & Doull's Toxicology: The Basic Science of Poisons*, 9th Edition. New York, McGraw Hill, 2018.

Lorke D. A new approach to practical acute toxicity testing. *Archives of Toxicology*. 54:275-87, 1983.

National Research Council, Division on Earth, Life Studies, Institute for Laboratory Animal Research, Board on Environmental Studies, Committee on Toxicity Testing, and Assessment of Environmental Agents. *Toxicity Testing in the 21st Century: A vision and a Strategy*. Washington, National Academies Press, 2007.

Riviere JE: *Chemical Food Safety*, Ames, IA, Iowa State Press—Blackwell Publishing, 2002.

Riviere JE: *Biological Concepts and Techniques in Toxicology: An Integrated Approach*. New York, Taylor Francis, 2006.

Trevan JW. The error of determination of toxicity. *Proceedings of the Royal Society of London. Series B, Containing Papers of a Biological Character*. 101:483-514,1927.

Williams FM, Rothe H, Barrett G, Chiodini A, Whyte J, Cronin MT, Monteiro-Riviere NA, Plautz J, Roper C, Westerhout J, Yang C, and Guy RH. Assessing the safety of cosmetic chemicals: consideration of a flux decision tree to predict dermally delivered systemic dose for comparison with oral Threshold of Toxicological Concern (TTC). Regulatory Toxicology and Pharmacology 76: 174-186, 2016.

Xu X, Riviere JE, Raza S, Millagaha Gedara NI, Ampadi Ramachandran R, Tell LA, Wyckoff GJ, and Jaberi-Douraki M. "*In-silico* approaches to assessing multiple high-level drug-drug and drug-disease adverse drug effects". *Expert Opinion on Drug Metabolism and Toxicology*. 20: 579-592, 2024.

Chapter 9
Cancer

> Almost all the world is natural chemicals, so it really makes you re-think everything. A cup of coffee is filled with chemicals. They've identified a thousand chemicals in a cup of coffee. But we only found 22 that have been tested in animal cancer tests out of this thousand. And of those, 17 are carcinogens. There are ten milligrams of known carcinogens in a cup of coffee and that's more carcinogens than you're likely to get from pesticide residues for a year! (Bruce Ames)

In the previous science-focused chapters, zero was often defined based on the inability to *experimentally* detect a chemical or observe an adverse biological effect. When the chemical or effect was not detected, one could reasonably conclude *nothing* would happen. This could occur under two conditions: (1) a *threshold* exists below which the assay was not capable of detecting the compound at that concentration, the LOD, or (2) the concentration was below a NOAEL-linked threshold, even if low concentrations of the chemical were present. This allows for the use of a D-R curve, which generally fits the observed experimental data, giving confidence to its credibility as a useful tool in risk assessment. This approach has worked for most drugs, consumer products as well as agricultural, and industrial chemicals.

Carcinogens and mutagens that interact directly with DNA represent a very different scenario, because any available data connecting a chemical exposure to an induction of cancer, is usually at exposure levels far greater than that encountered in foods or the environment. In the twentieth century, toxicology has been criticized for conducting high dose studies using very few doses to predict very low dose rare effects. As discussed in the previous chapter, when the effect can be observed with these doses and then disappear at lower doses in an experimental study (*e.g.*, the NOAEL), then a reasonable risk assessment can be made. In these cases, all data and predictions are within the inference space of the data in the study. The D-R curve covers both response and no-response exposures.

However, when dealing with effects that theoretically may occur at extremely low doses, scientists have difficulty extrapolating what the effect would be based on back-extrapolation from high-dose data, especially when the effect is one largely

---

Figure Use Attributions
Figs. 9.1, 9.2, 9.3, 9.4, and 9.5: Wikimedia – No restrictions
All other figures are either self-drawn, photographed or from author's own publications (attributed)

J. E. Riviere, *Zero – Much to Do About Nothing?*,
https://doi.org/10.1007/978-3-031-82998-7_9

based on dose extrapolations using theoretical mechanistic endpoints. At low doses, adaptive or hormetic responses may be induced, which may modulate any adverse effect, even a carcinogenic one.

Background on Chemical Carcinogenesis

Cancer is a disease that possesses a stigma unlike most other conditions. The term "cancer" is attributed to the fourth century BCE Greek physician *Hippocrates* (Fig. 9.1) who coined the word *karkinos* (crab) to describe the claw-like projections he observed in tumors. Note Hippocrates was also the author of the *Hippocratic Oath* that physicians to this day ascribe to.

Fig. 9.1 Hippocrates

In the mid-twentieth century, so little was known about cancer's origins or treatments that it was referred to as *The Great Darkness*. When a chemical was identified as being carcinogenic, it was believed that exposure must be avoided at all costs. This can be difficult to near impossible for many compounds, and today this fear still persists.

To estimate the minimal dose required to induce cancer, various statistical models incorporating different mechanistic assumptions are employed to extrapolate from high dose experimental animal data to a minimal dose that has the potential to induce cancer in humans. Research has been conducted on cancers induced by radiation, ultraviolet light, and biological organisms such as viruses. A significant amount of research conducted over the past decades has increased our understanding of the mechanisms of many carcinogens. This field is far beyond the focus of a book on zero and topics will be introduced solely to better understand how scientists estimate zero effects from the available experimental data.

Extrapolating to the lowest dose of a chemical that can induce cancer is a very difficult extrapolation because one cannot truly measure the lowest dose of a carcinogen *in vivo*. One must consider the previously discussed pharmacological obstacles of the ADME processes that any agent encounters upon entering the body before it interacts with its biological target. This is coupled with the possibility of adaptive hormetic responses being present at low doses that could ameliorate the response, or the need for exposure to a second chemical to complete the induction of cancer.

A carcinogen is any agent that promotes the development of cancer, defined as a *disease that causes normal cells to uncontrollably divide and proliferate by inducing a malignant transformation.* There are hundreds of types of cancers that have been identified, generally named based on the original cell-type they developed from. These cells can form tumors in solid tissues, escape the original tissue, and metastasize throughout the body (Fig. 9.2 showing tumors metastasized into the liver), or proliferate in fluid tissues like blood (*e.g.*, leukemia). Cancer is a genetic disease where genes that regulate cell growth and reproduction are damaged. Genetic changes capable of causing cancer include those random mutations that occur during normal cell division, those inherited from one's parents, or those that occur after being exposed to a carcinogenic agent.

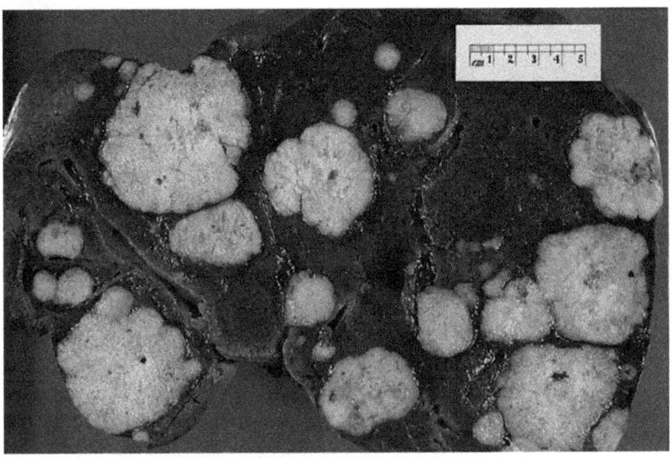

Fig. 9.2 Solid tumors that metastasized to the liver

Carcinogens include synthetic chemicals, naturally occurring substances, physical agents such as ionizing and non-ionizing radiation, and biologic agents such as viruses (human papillomavirus—HPV, hepatitis C), fungal products (aflatoxin), and even bacteria (*Helicobacter pylori* -indirect). Most carcinogens act by inducing mutations in DNA that disrupt a cell's normal processes for regulating growth, leading to uncontrolled cellular proliferation. This occurs when cellular DNA repair processes fail to identify DNA damage allowing the defect to be passed down to daughter cells, a phenomenon first identified as an inheritable mutation in the p53 tumor suppressor gene seen in *Li-Fraumeni Syndrome*. Multiple types of cancers occur even at a very young age because the ability to suppress carcinogenic mutations was blocked. Although first discovered as an inherited mutation, scientists now implicate chemical-induced damage to the p53 gene as causing a significant amount of human cancer (*e.g.,* breast, brain, colorectal, pancreatic, liver, lung, and ovarian cancers).

Cancerous cells under the microscope look different than normal cells as seen in Fig. 9.3. Cellular damage accumulates over time. This malignant transformation is typically a multi-step process during which the regulatory mechanisms within the cell are gradually disassembled allowing for uncontrollable cellular division.

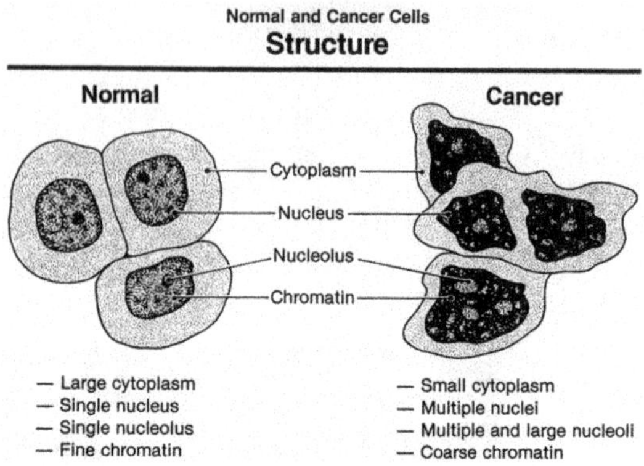

Fig. 9.3 Different microscopic appearances of normal and cancer cells

It is important to appreciate that chemical induced carcinogenesis is a muti-stage process involving the occurrence of several stages that ultimately result in a systemic disease. This makes identifying a specific triggering event very difficult. In its simplest form, two events are needed for a chemical to induce cancer: a mutational event (initiation) and proliferation of the mutated cell (promotion) to form a neoplasm (Fig. 9.4).

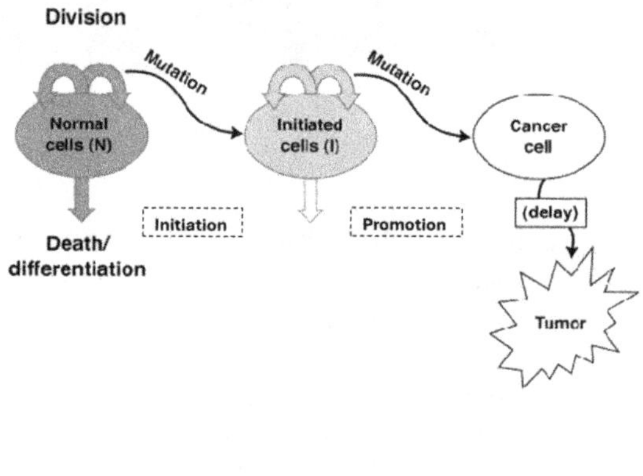

Fig. 9.4 Two-stage model of carcinogenesis

Hannahan and Weinberg further defined eight *Hallmarks of Cancer* that describe in greater detail the changes that occur to a normal cell's homeostasis that impart the abnormal growth and metastatic properties typified by a malignant neoplasm.

1. Inducing and sustaining cell proliferation
2. Resisting cell death or apoptosis

3. Inducing angiogenesis which enables metastasis
4. Enabling replicative immortality
5. Activating invasion and metastasis
6. Evading growth suppressors
7. Reprogramming energy metabolism
8. Evading immune destruction

Cancer is thus a process and not a single event. Carcinogens can be classified as either *genotoxic* or *nongenotoxic (epigenetic)*. In the two-stage mode of carcinogenesis, some carcinogens can only *initiate* the process by causing DNA damage that may remain dormant. Others *promote* the development of already initiated cells to form tumors or block suppressing the malignant growth. *Complete carcinogens* can do both. Many genotoxic compounds are complete carcinogens. As the reader can appreciate, the complexity of these models and biomolecular interactions is far beyond the scope of this chapter. Our focus is to get a better grasp on what zero means in this field.

Genotoxic carcinogens directly bind, damage, or interact with and modify DNA such that its normal functions are changed. This was the MOA for the earliest chemicals identified as carcinogens in the mid-twentieth century: benzo(a)pyrene (coal tar and soot), azo dyes, and some pesticides. Some carcinogens are *direct* acting in their original molecular form (*e.g.*, epoxides, imines, alkyl and sulphate esters, and mustard gas), while others are *indirect* as they require metabolic activation by enzymes in the body. Direct acting alkylating agents are generally carcinogenic to humans and other animals. Such highly reactive chemicals may also react with other molecules in the body and cellular environment, undergoing a competition between directly reacting with DNA *versus* other cellular components. In addition, they are the targets of detoxification enzymes and other cellular protective mechanisms. These processes modulate the actual exposed dose of alkylating agents available to interact with DNA. Most genotoxic carcinogens, such as aflatoxin, aromatic amines, nitrosamines, and polycyclic aromatic hydrocarbons, are indirect carcinogens requiring metabolic activation of the parent molecule.

Many chemicals are direct mutagens resulting in altered DNA adduct formation and sometimes altered transcription. However, in many cases, DNA adducts accumulate in tissues and do not immediately induce cancer as scavenger mechanisms remove them first. As discussed above, mutations in tumor suppressor genes may facilitate neoplastic transformation as would mutations that activate proto-oncogenes that cause neoplastic transformations.

The MOA of epigenetic carcinogens are more varied and include chemicals that alter cellular homeostasis such that tumors can develop. Tumors generally develop in tissues prone to developing tumors independent of chemical exposure. Chronic exposure to chemicals is generally required. MOAs of epigenetic carcinogens include repeated cytotoxicity, receptor mediated (CAR, PPARα, AhR) peroxisome proliferation, hormonal perturbations, and oxidative stress or altered methylation status causing DNA damage. A few inorganic metallic carcinogens also have epigenetic MOAs.

A complicating factor in assessing chemical carcinogenesis is that for many solid tumors to develop, it may take 10–40 years. This is termed the *latency period*. For blood cancers, the latency period may be as short as 2 years. Due to prolonged latency periods, identification of carcinogens associated with specific tumors may be quite challenging. Modern screening techniques focused on biomarkers of cellular transformation are starting to shorten this detection delay. Another issue with cancer is that there is a strong genetic basis for many cancers as well as interactions involving lifestyle and numerous environmental and dietary factors that make attribution to a specific chemical carcinogen difficult. Hormonal influences can also be dramatic for certain types of reproductive and endocrine cancers.

Cancers Bioassays

There are several approaches to determine if a compound is potentially carcinogenic. The carcinogenic potential of many compounds is hinted at by human epidemiological data that associates compound exposure to carcinogenesis. Human epidemiological studies can be broadly classified into three types:

- Cross-sectional studies that identify risk factors, exposure, and disease; but are not useful to establish a cause-effect relationship.
- Cohort or prospective studies that group individuals exposed and not exposed to a chemical agent which are then monitored for adverse effects.
- Case-controlled or retrospective studies where diseased individuals are matched with healthy individuals to identify factors that may be linked to the disease.

The design, strength, and validity of many of these studies vary greatly. Association data accurately identified benzo[*a*]pyrene in soot that caused scrotal tumors in chimney sweeps. Other studies are often confounded by environmental and health-style factors, for example not controlling for smoking or excessive alcohol consumption in study subjects. Many human studies use biomarkers of adverse effects rather than the actual presence of the disease, because they are less invasive to the study participant, are obtainable from existing medical records, or may indicate the potential to develop a disease that has not yet manifested. However, as discussed above, cancer is a multi-stage process, and some biomarkers may be too sensitive or indicative of a reversable disease process and not cancer. Most biomarkers are not chemical specific and could reflect exposure to another compound that was not controlled for in the study design. Biomarkers include markers of organ function as well as exposure to a carcinogen such as detecting DNA adducts or transformed tumor DNA. Finally, the statistical mantra that *correlation does not equal causation* must always be kept in mind when analyzing such data. Clinical trials for drugs and large well-designed representative cohort or case-controlled studies often can shed light on this issue.

Environmental outbreaks of cancer are generally detected by identifying "cancer clusters" in specific geographical or chemical exposure groups. These studies can identify true carcinogen exposures but may be confounded by clusters of unrelated cancers developing in groups due to many other factors. Many of these different cancers have different MOAs where one chemical cannot induce all the different types of cancer seen. Chemical exposures often occur in chemical mixtures, where carcinogens may be exposed at low levels, along with other chemicals that may either potentiate or inhibit carcinogenesis. Some workers hold that hormesis at low levels of exposure may induce protective physiological processes that decrease the potential for inducing cancer at low concentration exposures.

The most rapid and simple approach is to determine if the compound can cause DNA mutations, making it a candidate for being a genotoxic carcinogen. Tests include exposing test chemicals to bacterial and mammalian cells and using several endpoints to determine if DNA mutations have occurred (*e.g.*, Comet assay). These tests detect direct genotoxic carcinogens. For detecting indirect carcinogens, rat liver enzymatic homogenates are added to the assay allowing for metabolic activation of the parent chemical to its active carcinogenic metabolite. This is the basis of the widely used *Ames Test*. Various other human *in vitro* cell culture tests can be employed to detect altered chromosomes or direct DNA damage. These assays are required for drugs or consumer products being developed to remove genotoxic compounds from further testing. The detection of mutagenic compounds also helps to prevent those chemicals that could induce birth defects from entering the market. Despite the simplicity and utility of these rapid assays to detect genotoxic carcinogens, they cannot detect epigenetic carcinogens.

For the detection of skin carcinogens, for example when assessing cosmetic ingredient safety, *skin painting* studies are conducted in mice to determine if topical application of a test chemical is capable of inducing skin tumor formation. Different concentrations can be applied to assess the D-R relationship. A similar study design is used to study the two-stage model of tumor progression (Fig. 9.4), where an *initiator* induces the carcinogenic transformation while a *promoter* is then used to cause clonal expansion of the transformed cells. This allows chemicals to be tested for their ability to promote tumor growth in cells already transformed, for example by previous chemical or UV light exposure.

Briefly, I introduced laboratory animal toxicology studies in the last chapter that discussed how the NOAEL was determined. In that discussion, the caveat was if a compound could be a potential carcinogen, then lifetime exposure trials should be conducted to assess whether cancer could be detected. Carcinogenicity bioassays are performed as part of the drug approval process, despite most genotoxic drug candidates already being eliminated from study due to their mutagenic activity. In a typical experimental study whose design is a function of regulatory agency guidelines and the experimental questions being asked, mice and rats of both sexes are exposed in parallel studies. Two to three doses (low, middle, and high) of a compound plus a no-dose control group are generally employed. Three doses are required if part of the goal of the study is to generate data for a subsequent risk assessment where a dose-response curve will be generated.

The high dose is the dose of compound which allows animals to survive for the entire trial duration but does not induce significant adverse toxicological effects. In a cancer bioassay, the toxicology of the compound is generally well-known with the focus of the study being to determine its carcinogenic potential. This dose is determined from subchronic studies and termed the maximum tolerated dose (MTD). The MTD is also used in trials to detect teratogenic compounds where constant exposure does not cause adverse effects in the mothers, since the test subjects are the offspring of the exposed mothers. The reason for using MTD is to try and separate carcinogenic from other toxicological effects. An alternate approach is using the benchmark dose (BMD) discussed in the previous chapter, which is a lower dose determined from reviewing and modeling data that mirrors both the compound's mechanism of action as well as relevance to human exposure. The sophistication of these strategies has increased over the last few decades where MOA and computational models are employed to better estimate the potential a low dose exposure has on inducing cancer in animal trials. Statistical comparisons are then made between tumor or other cancer evidence in the control versus each treated group.

An experimental dilemma that often arises in these studies is *that a background signal of tumors always occurs in control animals where there was no chemical exposure. It can be exceedingly difficult to tease out these background spontaneous tumors from those induced by the chemical tested.* Some strains of laboratory mice and rats are especially prone to this problem. In some cases, there are precancerous changes that can occur with some chemicals inducing growth but not necessarily malignant transformation. The question then arises if a promoter agent {*e.g.*, chlorinated hydrocarbon pesticides such as chlordane, mirex, dichloro-diphenyl-trichloroethane (DDT), saccharin, or diethylstilbesterol (DES)}were also given to these animals, would tumors arise? There are species differences in chemical carcinogens, one example being the ability of tetrahydrofuran to only cause tumors in male rat kidneys but not humans.

Another dilemma is that for many carcinogens, tumors only occur at the highest-dose level of exposure in the bioassay. An example is TCDD (tetrachlorodibenzo-p-dioxin), which produces tumors in rats only at high dose levels. Is it then rational to back-extrapolate to a lower dose despite not detecting tumors at the low dose used in the bioassay?

Regulatory Bodies Defining Carcinogens

Carcinogens are identified by a few regulatory committees across different international jurisdictions using various criteria to classify such compounds. These include:

- International Agency for Research on Cancer (IARC) of the World Health Organization (WHO)
- Globally Harmonized System (GHS)
- European Union / European Commission panels
- National Occupational Health and Safety Commission (NOHSC)

- US National Toxicology Program (NTP) of the National Institute of Environmental Health Sciences (NIEHS)
- US Environmental Protection Agency (EPA)
- US National Institute of Occupational Safety and Health (NIOSH)
- American Conference of Governmental Industrial Hygienists (ACGIH)

Classifying a chemical as a carcinogen often makes it susceptible to regulations that strictly prohibit its exposure to humans. A major issue that plagues this field is that these agencies do not agree either on what data constitutes absolute evidence of carcinogenicity for a chemical and thus what chemicals are carcinogens. The IARC classifies potential carcinogens into four groups based on the strength and results of human epidemiological data and animal bioassay data.

1. Carcinogenic to humans: Human and animal data strong
2A. Probably carcinogenic to humans: Human data suggestive, animal data positive.
2B. Possibly carcinogenic to humans: Human data weak, animal data positive
3. Not classifiable as to its carcinogenicity to humans: Human and animal data inadequate
4. Agent is probably not carcinogenic to humans: Human and animal data negative

One can appreciate that there are a lot of subjective evaluations when classifying a substance using these criteria. Table 9.1 is a list of those substances IARC classifies as known carcinogens to humans.

Table 9.1 List of those substances IARC classifies as known carcinogens to humans

Acetaldehyde
Aflatoxins
4-Aminobiphenyl
Aristolochic acids
Arsenic
Asbestos
Azathioprine
Benzene
Benzidine dyes
Benzo[*a*]pyrene
Beryllium compounds
1,3-Butadiene
1,4-Butanediol
Dimethanesulfonate
Cadmium
Chlornapazine
Chlorambucil

(continued)

Table 9.1 (**continued**)

Chloromethylmethylether
Chromium VI
Ciclosporin
Cyclophosphamide
1,2-Dichloropropane
Diethylstilboestrol (DES)
Erionite
Ethylene oxide
Etoposide
Fluoro-edinite
Formaldehyde
Gallium arsenide
Lindane
Melphalan
Methoxsalen plus UVA
4,4′-Methylenebis(2-chloroaniline)
MOPP alkylating agents
Sulfur mustard gas
2-Naphthylamine
Nickel compounds
NNK–tobacco nitrosamine
N-Nitrosonornicotine
PCB-126
Pentachlorophenol
PFOA-
Polychlorinated biphenyls
Semustine
Silica dust
Tamoxifen[
TCDD
Thiotepa
Treosulfan
Trichloroethylene
o-toluidine
Vinyl chloride

Accessed May 18, 2024

The majority of compounds listed are also classified as carcinogens by other agencies. *It is important to realize that despite listing a compound as a known human carcinogen, there are no assertions made as to the dose of chemical required to induce cancer in humans.* Many of these compounds are carcinogenic because they are potent DNA alkylators inducing a mutation in the affected gene. Some are only carcinogenic when co-exposed with UV light, radiation, administered with other drugs, or exposed to a promoting agent. Others are exposure route dependent. For example, silica is listed due to inhalation of crystalline or silica dust in occupational scenarios. I am writing these words from a beach on the Outer Banks of North Carolina, which is primarily composed of silica sand. Should I be concerned because my feet are standing in the sand? Many scientists have criticized carcinogenicity rankings by the IARC. An example is the herbicide Roundup® whose active ingredient is glyphosate, which isn't a carcinogen, according to EPA, EFSA, the European Chemicals Agency, and regulators in Canada, New Zealand, Japan and Australia. Every regulatory body that has studied the compound has found it to be safe—with a single exception being the IARC which classifies it as 'probably carcinogenic to humans – Group 2A'.

As can be appreciated from the IARC classification scheme above, the two primary approaches these agencies use to classify a compound as being carcinogenic are laboratory animal bioassays discussed above coupled with strong human epidemiological evidence of a chemical association with cancer in humans. The animal trials are important because, as stressed by IARC *in the absence of adequate data on humans, it is biologically plausible and prudent to regard agents for which there is sufficient evidence of carcinogenicity in experimental animals as if they presented a carcinogenic risk to humans.*

The point to stress is that in many cases, if a compound is officially listed as a human carcinogen, all exposure to the compound is prohibited, leading to our discussion of what is zero exposure and is it relevant? This philosophy is consistent with the precautionary principle discussed in earlier chapters, where if there is any risk of serious harm, the best approach is avoidance. This leads to regulation not by concentrations known to induce an adverse effect but instead by simple detection. By now, the reader has been informed that with modern methods of analytical chemistry, extremely low concentrations below any ability to induce a biological effect can be detected.

Zero Dose Extrapolation

The primary issue in cancer risk assessment is thus how to extrapolate from available high dose data to the elusive zero dose. When the sigmoidal D-R relationship (Fig. 7.9) was discussed, the ED_{50} was introduced and became the LD_{50} in toxicological studies where 50% of individuals are dead after exposure. Some researchers have used the observed sigmoidal D-R relationship to try and extrapolate to lower effect levels such as LD_{10} (10%) or even LD_{01} (1%). The issue with accomplishing this rests purely on statistical issues arising from the small number of actual data points used to define the relationship. In cancer bioassays used for risk assessment, three doses and a control group are used. Despite sophisticated modeling approaches, the beginning or take-off point of the dose extrapolation is based on just three actual data points and reliance on the selection of a specific form (think shape) of the D-R model selected being valid.

We introduced the issues encountered in extrapolating beyond the observed data in the discussion of the data set in Fig. 4.9. To accurately extrapolate outside of the inference space defined by the actual observations, either additional data is required to expand the inference space or the model encompassing these lower concentrations must be valid at these concentrations. To further complicate the picture, data is variable (recall Figs. 4.5 and 4.7) making precise predictions difficult. As mentioned above, there is also a high background signal of spontaneous tumors in laboratory animals. All estimates are not point estimates, but rather confidence intervals that reflect the statistical uncertainties involved. Despite the apparent sophistication of modern cancer risk assessment models, much of the efforts to extrapolate cancer incidence at very low concentrations outside of the experimental data-defined window is fraught with uncertainty.

Estimating the median response of a data-sparse distribution, such as the $LD_{50,}$ is statistically robust. Estimating values at the tail of a statistical distribution, such as the normal distribution (Fig. 4.6), is difficult when little data is available. Many workers hold that three data points cannot be used to estimate what would be a 1 in a 100 event (LD_1) at a dose far below any tested. To counter this, different statistical distributions are used that are better at guesstimating (oops estimating) values at the extreme tails (e.g. Weibull distribution).

Figure 9.5 illustrates the vastly different predictions that occur using different statistical distributions and mechanistic cancer D-R models, with the probit and logit functions descriptive of our sigmoidal D-R relationship. The Weibull distribution is used to better estimate the lower concentration extrapolations. The other two extrapolations depicted below use the *multi-stage* carcinogenesis model or the most conservative *one-hit* model, where a single molecular event can transform a cell, to extrapolate back to doses needed to minimize the risk of cancer. There are many more mechanistic models now available that improve upon this extrapolation.

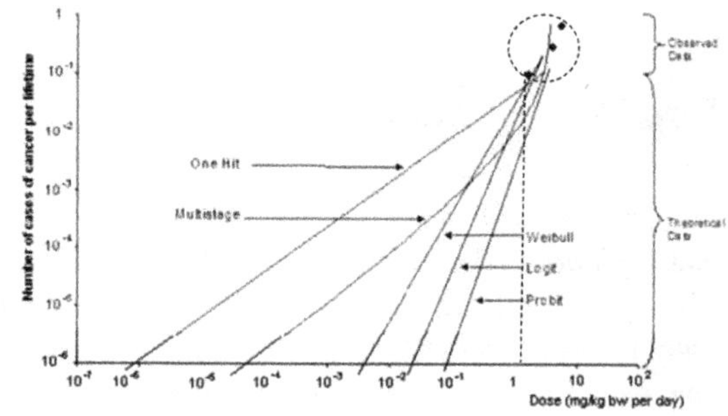

Fig. 9.5 Low dose extrapolation for a carcinogen. Dotted circle encompasses the inference space of actual data with the dotted line being the lowest dose actually tested. Everything lower than this is an extrapolation with estimates varying up to seven orders of magnitude from the lowest observed experimental data point

Today's models incorporate *in vivo* pharmacokinetic parameters, *in vitro* assays, and genomic analysis into *in silico* models considering the known MOA of the specific chemical being modeled, which defines the shape of an appropriate low dose extrapolation curve. The models incorporate principles of *Systems Biology,* which describe a toxic chemical's response from exposure to absorption, distribution, and finally interaction with the molecular target in a cell. These sophisticated models often contain data outside of the bioassay inference space in different test systems and alternative models to add some confidence to the extrapolations to lower exposure levels. The goal is to estimate the dose required to produce a specific amount of extra cancer cases in the population. Some model chemical concentrations (mass) and others convert to molarity (numbers) to reflect mechanistic molecular endpoints more accurately. Quantitative structural activity relationship (QSAR) analyses are conducted to relate a chemical's structure to its carcinogenic potential. This approach, coupled with machine learning algorithms, allows better definition of MOAs based on physical chemical parameters. This broadens the data available for input into more complex models because data within a chemical class can be used to predict effects based on similarity of MOAs.

Unlike the earlier models, which assumed a single molecule of a chemical could induce cancer, today's models attempt *to define a threshold for acceptable cancer risk*. There is great variability in this data and difficulty in defining the appropriate statistics needed to link the different scales of biology. Simply put, there are great gaps in our understanding of how to tie together these disparate processes measured under widely different laboratory conditions. Finally, how are sensitive populations considered in such risk analyses? Common reasons for increased sensitivity to toxic agents include deficiencies in liver or kidney function that impair the ability of the body to metabolize or excrete absorbed compounds as well as genetic factors that increase receptor responses.

There are an endless number of such models possible. They include both linear and non-linear models that take off from or extrapolate at the lower points of the observed data-based D-R curve (data enclosed in the dotted circle in Fig. 9.5), the so-called *point of departure*. Some investigators advocate that cancer bioassays should start at the low dose end of the D-R curve, and not the MTD, to better reflect actual life-time human exposures or exposomes. The *exposome* is a term that reflects the measure of all the exposures of an individual over a lifetime and how those exposures relate to health.

This could be at the BMD or NOAEL. These cancer risk assessment models are subjective and a function of both science and regulatory policy; the former which constantly changes over time as science progresses, while the later stays frozen in time until the glacial process of legislation changes the regulations.

Many exponential models "crash" when zero is approached for all the mathematical reasons mentioned in Chapter Four. Remember, when using cancer bioassays, there are only three actual observable datapoints used in the analysis, all occurring at high exposure doses. Fig. 9.5 clearly illustrates the dilemma of estimating a zero-cancer risk. In the context of our focus on defining zero, the lowest dose extrapolations are thus purely "*theoretical zeros*" defined by the data and assumptions used to derive the D-R models. This becomes a moving target whose use must be related to the context upon which it was originally defined.

Cancer Risk Assessment

In defining the nature of the models used to extrapolate from high dose animal data to define chemical exposure limits, risk assessors by necessity must define exposure and what is a tolerable risk of cancer occurring in the population. Animal cancer bioassays are lifetime exposure studies. Haber's Law is important in this context. Cancer risk estimates are generally based on the average lifetime daily dose, which is derived from the total cumulative exposure, *i.e.,* Haber's ($c \times t$). *Some risk assessors have defined an acceptable level of risk to a carcinogen as being the risk of one cancer per million people over a 70-year lifespan.* Researchers have suggested adding a tenfold safety factor for compounds where exposure to children is common.

Delaney Clause: Much of the regulatory zeal driving the quest for zero was driven by the *Delaney Clause*, which in 1958 amended the Food, Drug and Cosmetic Act (FDCA) of 1938. This *Food Additives Amendment* was promulgated when DES (diethylstilbestrol), a synthetic form of the female hormone estrogen, was prescribed to pregnant women from the 1940s to early 1970s to prevent miscarriages (later shown not effective) and other pregnancy-related complications. Daughters of DES-treated mothers were later shown to have an increased risk of developing a rare type of vaginal and cervical cancer (clear cell adenocarcinoma) with an incidence of 1.5 cases for every 1000 women exposed. This exposure was based on high dose therapeutic DES doses. However, DES was also used as a growth promotion agent for cattle and thus could appear as a residue in beef at very low concentrations. At the same time, concerns were raised about the potential adverse effects of widespread exposure to saccharin and nitrites in foods. This fear influenced Congress to pass the *Delaney Clause,* which disallowed adding any potential carcinogenic additive to food.

This ban was originally based on the belief at the time that even one molecule of a carcinogen could cause cancer in humans, a concept largely influenced by theories of radiation-induced cancer. *Thresholds were not allowed in carcinogenic risk assessment. This is no longer considered valid.* Carcinogenesis is a multi-stage transformation. The ADME processes of absorption, distribution, metabolism, and elimination combined with cellular defense and repair mechanisms, makes the possibility that individual molecules of chemicals could induce cancer next to impossible. Note that unlike chemical exposure, there are no true barriers protecting DNA from radiation damage.

When Congress and the regulatory community realized that this earlier thinking may have had flaws, in a 1968 amendment to the Delaney Clause—*the DES Proviso* linked the allowable limit to an analytical method. Thus, carcinogenic drugs could be used in animals destined for food if one could not detect residues in the tissues. At this time, analytical chemistry was such that part-per-million detection of chemicals was considered an extremely sensitive assay. Now we can detect a million times less chemical, and at levels that were never anticipated when these laws were passed.

Beginning in the early 1970s, the FDA began working on a sensitivity-of-method approach to regulating drug residues in food, which was ultimately published in

1987. This approach, currently in use, defines a threshold residue level below which food is deemed "safe" even if analytical methods can detect a residue. The residue level is set to a life-time risk of cancer of 1 in 1,000,000. This risk of cancer is determined by using the types of animal bioassays, epidemiological and statistical modeling studies presented above.

The scientific debate, which lasted for decades, is over, and almost everyone accepts that detection below an established tolerance adds an immeasurable level of risk to the food supply. *Analytical detection was finally uncoupled from a potential to cause a biological effect*. The FDA established a *de minimis* risk level of 1 ppm for the trace contaminants, residues, or impurities found in an ingredient. The *de minimis* level reflects the minimal risk that a component of an ingredient could cause harm. [Note that this is not to be confused with FDA's other *de minimis* rule (Section 321) which allows the clearance of economically low-value shipments without payment of duties or taxes.]

Yet even today, it is detection at these Lilliputian levels that causes the media hysteria and a tremendous amount of unfounded anxiety in the public's psyche. This "no-tolerance" philosophy has removed many chemicals from our food supply and fostered the development of safer, more effective therapeutics and less potentially toxic food additives.

Conclusion

This chapter on carcinogens completes the discussion necessary to grasp a better handle on the meaning of zero in toxicology. In the chapter on non-carcinogenic toxins, extrapolation of D-R curves to the NOAEL or BMD was relatively straightforward, because it is possible to define a threshold exposure where no adverse effects were observable. This serves as a reasonable operational definition of zero or no effect.

For carcinogens, this is much more difficult because the disease is a multistage process, and the potential risk of cancer is assessed at concentrations far below where experimental data is collected. Cancer is rightly associated with anxiety in people, especially when the proposed route of exposure in food or water is beyond their control. Confusion reigns when new scientific findings accentuate certain risks and eliminate others. The continued rapid advances in analytical chemistry result in chemicals that are identified as known carcinogens being detected at levels far below the ability to produce any biological effect. Thus, there may be a zero effect in biology coupled with a non-zero concentration in analytical chemistry.

Animal bioassay data and human epidemiological extrapolations are based on life-time exposure. Does that suggest that a single exposure to a carcinogenic compound should be avoided when data shows that chronic exposures are required for an effect? In many cases, this is the source of public concern even when exposure is below the threshold required for a biological effect. Does constant exposure to a carcinogen that can only promote the development of initiated cells have a threshold? Is chronic exposure required for such an effect?

In the fields of pharmacology where the dosages required to be safe and effective are defined in clinical trials or in toxicology where exposure to a chemical in food, consumer products, cosmetics, water, or the environment can be shown to have minimal risk of adverse effects, thresholds are the basis for operationally defining zero effect. Here is where the confusion arises, as discussed in the chapter on chemistry, a chemical may be quantified or detected at levels below the toxicological threshold. For chemicals, such as carcinogens, where thresholds may be fuzzier than for non-carcinogens, is an analytical zero necessary? This is further confounded by the fact that laws generally regulate carcinogens as all being the same across chemical classes when science clearly shows that they are different and often work by different mechanisms.

It is these ever-shifting sands of analytical and biological definitions of zero that add confusion to arriving at a clear definition. This will be explored further in the field that has had the greatest impact on human interpretation of zero, which is food safety (Chap. 10) as well in the human and equine sport world (Chap. 11) where detection of a banned substance, even one without athletic enhancing effects, leads to disqualification. Zero sure had a knack of causing chaos.

Selected Readings

Abelson P. Testing for carcinogens with rodents. *Science,* 249: 1357, 1990.

Ames B. Identifying environmental chemicals causing mutations and cancer. *Science,* 204: 587-593, 1979.

Beal SL. Ways to fit a pharmacokineitc model with some data below the quantification limit. *Journal of Pharmacokinetics and Pharmacodynamics,* 28: 481-504, 2001.

Boobis AR, Flari V, Gosling JP, Hart AD, Craig PS, Rushton L, and Idahosa-Taylor E. Interpretation of the margin of exposure for genotoxic carcinogens—elicitation of expert knowledge about the form of the dose response curve at human relevant exposures. *Food and Chemical Toxicology,* 57: 106-18, 2013.

Hanahan D, and Weinberg RA. Hallmarks of cancer: the next generation. *Cell.* 144:646-674, 2011.

Hilton GM, Corvi R, Luijten M, Mehta J, and Wolf DC. Towards achieving a modern science-based paradigm for agrochemical carcinogenicity assessment. *Regulatory Toxicology Pharmacology.* 137:105301, 2023.

Honma M. An assessment of mutagenicity of chemical substances by (quantitative) structure–activity relationship. *Genes and Environment.* 42:23, 2020

Klassen C. *Casarett & Doull's Toxicology: The Basic Science of Poisons*, 9th Edition. New York, McGraw Hill, 2018.

Riviere JE. *Biological Concepts and Techniques in Toxicology.* New York, Taylor & Francis, 2006.

Rozman KK, Kerecsen L, Viluksela M K, Österle D, Deml E, Viluksela M, Stahl BU, Greim H, and Doull J. A Toxicologist's View of Cancer Risk Assessment. *Drug Metabolism Reviews, 28*: 29–52, 1996.

Weinberg RA. *The Biology of Cancer, 2nd Ed.,* New York,Garland, 2013.

Zhu Y, and Costa M. Metals and molecular carcinogenesis. *Carcinogenesis.* 41:1161-72. 2020.

Chapter 10
Food Safety

> Everything I eat has been proved by some doctor or other to be a deadly poison, and everything I don't eat has been proved to be indispensable to life.....But I go marching on. (George Bernard Shaw)

We have reached a point where we can consider how the science-based definitions for zero such as the NOAEL or LOD, are integrated into the legal definition of how much of a chemical humans should be allowed to be exposed to with no risk of adverse effects. *This is the zero that we will focus on.* Toxicologists and public health officials are rightfully concerned with human exposure to potentially toxic chemicals that is constant and out of an individual's control. These exposure routes include food, water, and air. Occupational and environmental exposures are usually known and strictly regulated once the hazard is identified by federal regulatory agencies such as ATSDR, EPA, and OSHA as well as state public health authorities. Food is the exposure route where most of the issues related to analytical detection, toxicological endpoints, and regulatory definitions interact. Food safety is also the driving force behind many news stories where detection of hazardous chemicals in food sounds an alarm.

Chemical food safety encompasses the entire production cycle of food from farm to table (Fig. 10.1). The diversity, breadth, and regulatory control over this essential human endeavor is often underestimated. Chemicals can be introduced at the farm level, as animal drugs or crop pesticides; or inadvertently as environmental or water contaminants; at the production-level when preservatives, colorings, flavorings, and other food additives are added to processed foods; or during transportation and warehouse storage where packaging may degrade, or spoilage or mold growth could occur and release toxins. The result is incorporation of the food product into a meal served at dinner. There are multiple points in this process where potentially harmful chemicals may become adulterants in the final food product. There is no single regulatory agency responsible for all food from farm to table. In the author's experience, it is at the interfaces where these issues arise.

Figure Use Attributions
Figs. 10.3, 10.4, and 10.6: Riviere JE. *Comparative Pharmacokinetics: Principles, Techniques and Applications,2nd Ed*, Ames, IA, Wiley-Blackwell, 2011. c Jim Riviere
All other figures are either self-drawn, photographed or from author's own publications (attributed)

J. E. Riviere, *Zero – Much to Do About Nothing?*,
https://doi.org/10.1007/978-3-031-82998-7_10

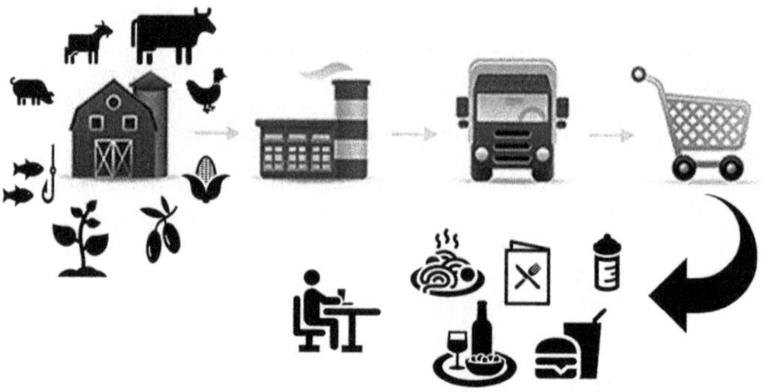

Fig. 10.1 The journey from farm to table

We will focus on chronic chemical hazards because most food poisoning scenarios are acute and a result of bacteria or high level irritant chemical absorption. This makes people sick shortly after consuming contaminated food. Illness confirms exposure making a discussion of zero irrelevant for this scenario. In most cases, this acute illness is the body's natural defense mechanism working to rapidly expel the tainted food from the body through vomiting or diarrhea. This is not the case for trace level exposure to chemical contaminants.

Determination of Safe Levels for Human Consumption in Meat and Milk

The starting or departure point for all quantitative metrics of food safety is the animal-derived toxicological endpoints such as the NOAEL or RfD, and the analytical chemistry sensitivity thresholds such as LOD and LOQ. As we discussed in the last chapter, if the chemical is a potential carcinogen, the *DES proviso* shifts the tolerance to some low level of analytical detection. The task is converting this toxicological threshold to an allowable or so-called *safe level* in food, which becomes food product dependent. This conversion is based on assuming *lifetime exposure to the chemical at this safe level* and on *how much of the food that contains this chemical is consumed.*

The US FDA is required by statute to determine whether edible products of animal origin (*e.g.*, meat, milk, eggs, or honey) from production animals treated with animal drugs are safe for human consumption. Regulations that implement these requirements are found in Title 21 Parts 514 and 556 of the US Code of Federal Regulations. This process of animal drug approval is well-defined and nicely illustrates the factors that must be taken into consideration when defining safe levels for human exposure. With most animal drugs, residues may be detected in the food product but must be below the safe level defined by the relevant regulator. These are termed *tolerance* in the US and *maximum residue level* (MRL) in many other countries, including the European Union as well as the international Codex Alimentarius commission (CODEX), a joint program of the United Nation's Food and Agriculture Organization (FAO) and World Health Organization (WHO) that sets recommendations for global food standards.

In food animal production, the interval between the time of the last administration of an animal drug and the time it takes for drug or chemical to deplete below tolerance concentrations, indicating that the animal can be safely slaughtered for food, is termed the *withdrawal time*. In dairy cows, milk can only be consumed after a *milk discard-time* has elapsed. These are legal dictates that the animal producer must follow before product is released to the market for human consumption. The approach to determine these times is fully discussed below.

The first step in this process is defining the *Acceptable Daily Intake* (ADI). The ADI is the daily intake of total residues that over the life of an individual, is below the NOAEL and thus should not result in adverse effects or harm to the health of the consumer.

$$\text{ADI} = \text{NOAEL}\left(\text{mg / kg Body Weight / day}\right) / \text{Safety Factor.}$$

Food safety regulators are extremely conservative when developing safety metrics, acknowledging that extrapolation from animal studies to humans and estimating population responses to chemical exposure is a complex process. Safety or uncertainty factors are used to protect against underestimating toxicity. Their values are dependent upon the regulatory jurisdiction involved and the strength of the underlying toxicology data. Standard safety factors of 10X account for animal to human differences and another 10X account for exposure to particularly sensitive individuals such as children. *This yields a minimal safety factor of 100X.* If the compound is a teratogen potentially causing birth defects, another 10X may be applied. The EPA adds an additional 10X factor to account for deficiencies in study design. If the underlying toxicological study were of shorter duration, a factor may be added to cover longer exposures. For many food additives, regulated by the FDA and USDA, a 2000X factor may end up be employed. *The use of safety or uncertainty factors in calculating ADI significantly reduces the NOAEL, which recall is the only value derived from the experimental data and represents the no observed biological effect!*

For antimicrobial drugs, microbiological endpoints are also considered when establishing a microbiological ADI. These include an increase of the population(s) of antibacterial resistant bacteria seen with drug exposure, and/or the disruption of the microbial colonization barrier. This value is then compared to the normal ADI and the lowest value selected.

The next step is to determine the *safe concentration (SC)*. This is the amount of total residue of compound allocated to each edible tissue that can be safely consumed every day for life without exposing the consumer to residues more than the ADI.

$$\text{Safe Concentration} = \frac{\text{ADI} \times \text{Human Body Weight}\,(60\,\text{kg})}{\text{Food Consumption Value}}$$

The food consumption value represents how much of an animal's tissue would be consumed daily by an individual. These values in the US, expressed as grams of edible tissue consumed per day, currently are:

Muscle	300
Liver	100
Kidney	50
Fat or Skin/Fat	50
Eggs	100
Honey	20
Milk	1.5 L

Calculation of each edible tissue's SC reflects the partitioning of the ADI across edible tissues. In the US, the process assumes that a person daily consumes full portions of milk, eggs, and honey in addition to a full portion of one of these edible tissues. Other regulatory jurisdictions assume consumption of every tissue and not just one. However, regulations generally require that the total maximum daily intake (TDMI) calculated across these tissues and food products does not exceed the ADI. For some products not used in dairy cows or laying hens, ADI can be allocated to edible muscle, liver, kidney, or fat without partitioning. As the reader can appreciate, there are many assumptions that are incorporated into the determination of safe concentration. Importantly, this safe concentration is based on consumption of this quantity of meat and milk and eggs every day for life.

Determination of the Withdrawal Time

We now have defined the safe concentration as the target concentration for specific food products. Our goal is to determine what the threshold concentrations are across different tissues because the ADI is partitioned across various tissues. There are several issues that influence this determination. Figure 10.2 illustrates that the decay of drug residues occurs at different time points across tissues even if the tolerance is the same. {Note that in pharmacokinetic plots of drug or chemical decay, the depletion curves are straight lines when plotted as log concentration versus time (semi-log graph) because the equations governing their decay are first order processes—see Chap. 4}. Note, there is no zero on the x axis!

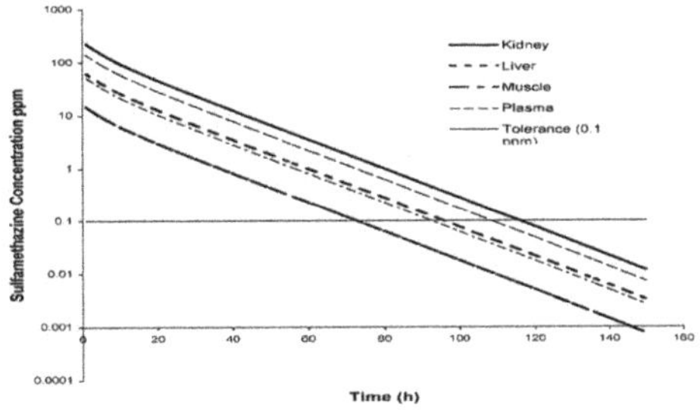

Fig. 10.2 Depletion of the antimicrobial drug sulfamethazine in different tissues down to the tolerance of 0.1 ppm

A target tissue must be defined to monitor residues at slaughter. The *target tissue* is normally the last tissue where residues *deplete* to their safe concentration. In the example above, this would be the kidney. If a drug is metabolized, the safe concentration reflects the depletion of total residues, not just the parent drug. In these cases, a *marker residue* is selected as the residue, whose concentration is in a known ratio to the concentration of total chemical residue in an edible tissue, allowing it to be monitored for the presence of residues. It can be the parent drug or one of its metabolites. This is experimentally determined in the total residue study, which utilizes radiolabeled compounds and monitors decay of all drug products in major tissues in healthy animals. The point where the total residues deplete to the safe concentration determines the value of the tolerance. This tolerance for the specific marker residue is then used in new animal trials to estimate withdrawal time when new products and formulations of a commercial drug are developed. The assumption is that the ratio of marker residue to total residues would still hold making the tolerance the target concentration to achieve.

The tolerance is defined as the concentration of the marker residue at the withdrawal time. Figure 10.3 illustrates this relationship. The absence of marker residue detection in target tissue above the legal tolerance confirms that the total residue concentrations in all other tissues is less than their respective safe concentrations. Tolerance is the maximum concentration of the marker residue that can legally remain in the target tissue of the treated animal. If the marker residue is less than the tolerance value, then total residues are less than the safe concentrations in all tissues. Monitoring the marker residue in the target tissue ensures the safety of the carcass. Note that if a drug is not metabolized, the parent drug represents total residues and is the marker residue.

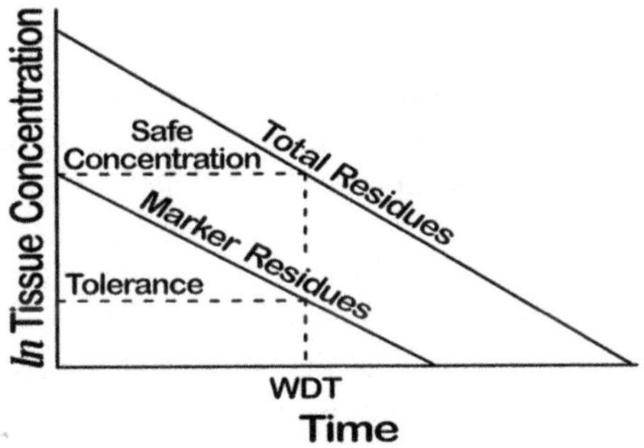

Fig. 10.3 Relationship between total residues and safe concentration, and marker residues to tolerance

In our illustrations above, we are only looking at a single depletion curve that usually represents the upper 99% of the population (1 out of 100) estimated with a statistical confidence of 95% from the actual experimental observations. Figure 10.4 illustrates this distribution for a hypothetical drug. It is the right-most upper limit of the confidence interval that determines the tolerance. There are very specific statistical methods and experimental designs (*e.g.*, TLM—tolerance limit method) required to calculate this value. *Operationally we define zero residues as the drug concentration in animal tissue or product below the legally established tolerance.*

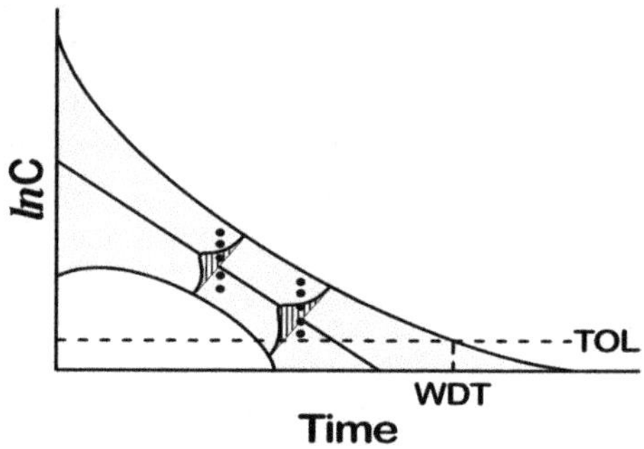

Fig. 10.4 Determination of tolerance as the 99% percentile estimated with a 95% confidence interval

Another scenario is where the tolerance based on toxicological data is below the limit of detection of the assay for the drug. For non-drug chemicals, such as pesticides, where the NOAEL or BMD is established and then safe concentrations are calculated, the LOD could exceed the calculated tolerance. When new toxicology data are generated, the revised tolerance may be below the LOD. This is a rare scenario today for pharmaceuticals due to the high sensitivity of modern analytical techniques. For drugs, the sponsor is now required to develop an assay that can detect drug at the tolerance concentration. However, for pesticides and environmental contaminants, this may be pertinent.

In some cases, the residue is tightly bound to tissues after exposure. Toxicological tests (relay toxicology studies) then assess whether this bond can be broken after digestion and release free drug back into the body. For drugs injected by needle, higher residue levels may be found trapped at the drug injection site and do not reflect other tissue levels. Similarly, residue concentrations differ among different muscles. Residue concentration can vary depending on tissue sampling protocols. Drug and chemical disposition may be different between young animals (*e.g.*, calves), adults and lactating mothers (*e.g.*, dairy cows), making residue tolerances different in these different production classes of animals. Production classes of animals, such as calves *versus* adults, or broiler *versus* layer chickens, may have different tolerances in different jurisdictions depending upon the local culture's food consumption preferences and restrictions. There are many different scenarios that arise from differences in pharmacology, as well as on the relationship between food safety toxicological thresholds and limits of analytical chemistry, all of which can result in some complex residue management issues.

In the US, residues are monitored in a variety of different sampling programs by the US National Residue Program (NRP). Residues are often scored as:

- *Zero:* Below LOD
- *Trace*: Between LOD and LOQ
- *Detectable*: Between LOQ and Tolerance
- *Violative*: Greater than the Tolerance

For the NRP, official assays of contaminants must adopt an analytical performance metric termed the *Minimum Level of Applicability (MLA), which is defined as the lowest level at which an analytical method has been successfully validated for a residue in each matrix (meat or poultry).* The FSIS will usually not report or act on any analytical results below the applicable MLA. This essentially is the LOD officially defined for an approved method whose data is reported by NRP.

There are differences between the approaches used by various regulatory agencies across the globe which result in different tolerances and MRLs for the same species, tissue and drug (Table 10.1; Fig. 10.5). These include:

- Different statistical limits on defining the lower confidence interval for threshold or MRL, that is the 95th rather than 99th percentile.
- Different safety factors applied to the NOAEL in determining the ADI.
- Different food consumption values.
- Use of a non-parametric rather than parametric approach (Fig. 10.5) where the withdrawal time is based on the point where drug falls below MRL or detectability in a series of samples collected at increasing intervals after cessation of drug exposure. There is no attempt to fit the data to a defined mathematical depletion profile.
- Selection of different official assays to monitor for residues.
- Historical date when the drug or tolerance was officially codified, as this impacts the level of analytical chemistry that existed at that point.
- Low tolerance or MRL was established as a barrier to trade.

As can be appreciated by looking at this figure and table, there is great variability in official tolerances and MRLs. Trading partners often share the same thresholds, and several countries are active in global harmonization efforts to achieve consistent endpoints. Countries have their own tolerance to risk and often establish default *de minimis* MRL levels for importing compounds without a specific established MRL, termed import tolerances (see section on quantitative toxicological endpoints in Chap. 8). For example, Japan sets four classifications for importing agricultural chemicals on their list:

- Chemicals subject to import residue limits on positive list system
- Chemicals known to pose significant risk and not subject to any limitation
- Prohibited chemicals that should not be detectable in food (< LOD)
- Chemicals for which no specific limit has been set and thus are subject to the uniform limit of 0.01 ppm.

There are numerous variations to such systems across global jurisdictions. They aim to reduce the risk for the human consumer to essentially zero risk for adverse effects.

Table 10.1 MRLs across the world

		Ampicillin			Ceftiofur			Fenbendazole		
Market	Species	Kidney	Liver	Muscle	Kidney	Liver	Muscle	Kidney	Liver	Muscle
Australia	Cattle				2000	2000	100	100	100	100
Brazil	Cattle	50	50	50	6000	2000	1000	100	500	100
Canada	Cattle	10	10	10	6000	2000	1000	200	450	100
China	Cattle	50	50	50	6000	2000	1000	100	500	100
CODEX	Cattle				6000	2000	1000	100	500	100
EU	Cattle	50	50	50	6000	2000	1000	50	500	50
Japan	Cattle	30	40	30	6000	2000	1000	100	500	100
New Zealand	Cattle	50	50	50	6000	2000	1000	50	500	50
Russia	Cattle	50	50	50	6000	2000	1000			
United States	Cattle	10	10	10	400	2000	1000		800	400
Brazil	Swine	50	50	50	6000	2000	1000	100	500	100
Canada	Swine	10	10	10	5000	2000	1000	2000	4500	1000
China	Swine	50	50	50	6000	2000	1000	100	500	100
CODEX	Swine				6000	2000	1000	100	500	100
EU	Swine	50	50	50	6000	2000	1000	50	500	50
Japan	Swine	9	60	60	6000	2000	1000	100	500	100
New Zealand	Swine	50	50	50	6000	2000	1000	50	500	50
Russia	Swine	50	50	50	6000	2000	1000			
United States	Swine	10	10	10	250	3000	2000		3200	2000

All values are in ppb. Blank cells reflect no available value. Data obtained July, 2024 from web search of public global databases curated by Drs. M Jaberi-Douraki and N Zad

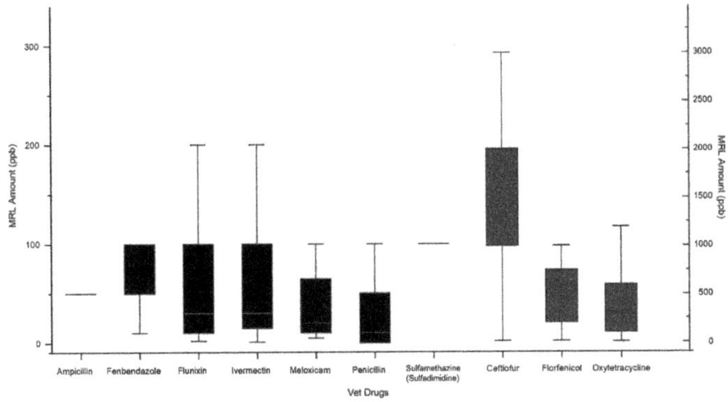

Fig. 10.5 Distribution of the range of MRLs of 10 drugs across multiple global jurisdictions and species ranging from 0 to 3500 ppb. (Zad et al., Food and Chemical Toxicology, 2023). Right three drugs use right concentration scale

Some regulatory jurisdictions set abnormally low thresholds for specific drugs as trade barriers to block import of food into their countries if residues exceed this artificial threshold. These may be related to specific species, are often negotiated in trade agreements, and do not follow international standards such as CODEX. All calculations, except for specific barriers to trade, consider concentrations determined using the ADI and safe concentration approach presented above, or simply use LOD as the default barrier to no exposure. Economic considerations may come into play to ensure that even detection of a chemical residue doesn't trigger a public reaction with resulting consumer avoidance of the product. *This approach, as will be discussed in the next section, has nothing to do with science-based food safety activity but rather manages consumer reaction to the detection of chemicals in food.*

Our understanding of chemical residues in animals is largely based on animal drug approval trials. Our understanding of chemical toxicity in humans and its predictability from laboratory animal studies is based on the clinical trials used for human drug approvals. This knowledge allows one to make reasonable risk assessments for animals that have been exposed accidentally to chemicals on the farm based on applying the same principles used to establish withdrawal times.

The approach is straightforward. The tolerance for the suspect chemical (*e.g.*, a pesticide) is determined. This becomes the target concentration. If there is no established MRL, a *de minimis* level may be selected (0.5–10 ppb) or the LOD or LOQ is used depending on the nature of the chemical contaminant. The value selected is based on subjective judgements of the potential toxicity of the compound and how much data is available to make the prediction. The pharmacokinetics of the pesticide is examined. The important point is to ensure that the data available actually covers the time frame of the predictions to be made (recall discussion of inference space in Fig. 4.9). Often, a study might have monitored high concentrations related to efficacy rather than to trace concentrations of residue concerns. This is illustrated in Fig. 10.6 that depicts two sets of experiments for the same drug, one over 10 h and the other over 80 h.

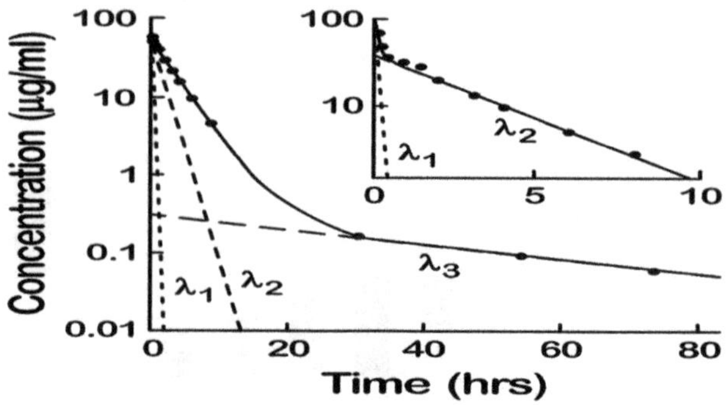

Fig. 10.6 Concentration versus time profile for a drug of experimental durations of 10 (insert) versus 80 h. λ is the slope of one of the linear portions of a multi-phasic plot. Note that concentration axis of µg/ml is same as ppm

This figure nicely illustrates many of the issues encountered when extrapolating data from high to lower concentrations. If only the insert data is available, extrapolating time to 0.01 µg/ml (0.01 ppm) would have the intercept of λ_2 with the x axis of about 12 hrs. However, longer observations show this chemical exhibiting a slower depletion phase (λ_3), which has concentrations still exceeding 0.1 ppm at 80 hrs. It could take a week before the target of 0.01 ppm is reached. Collecting a few samples can often validate the model under field conditions. Recall that drug and chemical decay is usually a first order process, a value of zero on the x axis can never be achieved as the logarithm of zero is not defined.

This *nasty issue of zero in such computations* is the primary problem with using the type of regression equations that have been discussed in risk assessment. One cannot simply input zero as the target. In practical terms, if the tolerance is not known, some used LOQ or LOQ/2 as the target concentration. Others propose LOD, however, the limit of detection often has a large statistical uncertainty. The compromise of (LOD + LOQ) / 2 has been suggested. Some input an arbitrary low value simply to ensure that the calculation doesn't crash the computer. However, these *ad hoc* approaches will distort the variance structure of the pharmacokinetic model. In the regression analysis, this value could then be given less weight in the model's solution.

As an example, imagine a study is conducted over 4 weeks and concentrations are below the LOD at 3 and 4 weeks. From a food safety perspective, one knows that no drug is present at 3 and 4 weeks. However, some regulations like the TLM require a regression analysis to be performed. If one ignores the LOD values, the earlier time points might derive an equation that does not reach LOD at 3 weeks. In this case, the analysis ignores the absence of detecting drug, which is a real observation and should be incorporated into the assessment. There are statistical methods (*e.g.*, censored multivariate regression, reduced weighting) that attempt to deal with this issue, but all have drawbacks. The simplest approach is usually to use LOQ/2. The alternate approach is the non-parametric one that would set the withdrawal time at 3 weeks when drug is no longer detected.

Take Home of Food Safety Thresholds

In this brief overview of how tolerance is established, a few points should be stressed. The first is that actual data from controlled experimental trials in animals is only the beginning of determining allowable dietary intake levels. As discussed in Chap. 8, where the tools of risk assessment were introduced, various laboratory animals and alternative *in vitro* tests have been developed to provide a better understanding of the MOA and what type of risks can be tolerated. The most common scenario is that an animal trial conducted in multiple species clearly establishes a NOAEL based on the data collected. *It is important that one realizes that this experimentally determined NOAEL (or BMD in other regulatory scenarios) is an indication of the absence of toxicity in the presence of chemical exposure.* This safe value is the *departure point* from which safe concentrations and tolerances that are monitored in the food supply will be determined. Due to the incorporation of uncertainty factors, the safe concentration may be hundreds to a thousand times lower than the NOAEL determined from the animal trials. *Is detection really a concern?*

The second point is that these safety calculations are based on lifetime consumption of food products (*e.g.*, meat, milk, or eggs) that contain this threshold concentration. Very few individuals in the population eat such a consistent diet for life or obtain their food from the same sources where contamination was present. *It is illogical to think that short-term exposure to a trace amount of chemical in food at concentrations below the established threshold has any potential for adverse effects! Detectable levels below tolerance really are zero risk.*

The third point is that standards across different regulatory agencies within a country and across global jurisdictions generally follow similar approaches to establish safe concentrations, but use different assumptions, daily food intake values, and uncertainty safety factors. There is often little correlation between tolerances and MRLs from different jurisdictions for the same compounds and species of animals. This is clearly seen in the range of values across jurisdictions for the same compounds in Fig. 10.5.

Additional confusion arises if the chemical is identified as a potential carcinogen. What agency identified it and under what exposure conditions? How does one interpret such trials when tumors are only observed at high doses in the laboratory animal trials, yet chemicals are regulated at close to analytical detection limits? Should a short-term exposure to trace levels of a potential carcinogen be of concern considering the pharmacokinetic processes present in the body to dampen absorption and distribution and increase elimination; as well as by action of built-in protective systems that reverse or eliminate damage in many cells once detected.

A final factor, often encountered in all these discussions on the legal definition of zero, is the historical context behind their determination. The sensitivity of analytical assays has greatly increased. The design of laboratory animal trials, coupled to modern computational and alternative model toxicological systems, has improved our ability to define the MOA and toxicological profile of many compounds. This allows regulators such as the EPA to often use abbreviated studies when the properties of a new compound can be predicted from similar compounds, a technique termed *read-across toxicology*. These results can justify the use of *de minimis* or *limit thresholds* such as 0.01 ppm for compounds related to others already studied. Resources can then be applied to those compounds for which data are lacking.

In the realm of food safety, zero must be clearly defined. Is one thinking about zero compound being present based on analytical assays (concentration $<$ LOD or LOQ) or is zero risk of adverse effects (concentration $<$ tolerance or MRL) being discussed? Has a value been substituted for zero in an equation simply to avoid indeterminate mathematical calculations?

Microbial Food Safety

Our discussion on food safety has focused on chemical residues, yet the major cause of food safety incidents globally is microbial contamination. Depending on the food type, this typically is from pathogenic bacteria including *E Coli, Salmonella sp., Listeria sp.,* and *Campylobacteria* etc. Symptoms can range from mild gastrointestinal discomfort to death. Microbial contaminants can be detected in plant and animal food samples by either culturing them or assaying for genetic or protein markers indicating their presence. There are numerous caveats in their detection since most involve amplification of the microbe or its components using classic bacterial culturing to molecular biological techniques such as the polymerase chain reaction (PCR).

For an individual eating contaminated food, there must be a sufficient dose of infectious microorganisms present, a term called the *minimal infective dose* (MID), to cause disease. For example, approximately 100,000 *Salmonella* organisms are needed to cause a gastrointestinal disease, while less are needed for *Listeria*. This is an over-simplification since the MID is very dependent upon the health of the consumer and the genetics of the infecting bacteria which affect properties such as:

- The virulence and pathogenicity of the specific microbial strain present
  - Infectivity including adherence factors, motility, and ability to penetrate host barriers
  - Lethality such as capsule formation and toxin production
- The status of the individual's immune system
- Overall health of the individual
- Gastrointestinal tract factors such as gastric acidity
  - Higher gastric acid pH after used of antacids is less lethal to ingested bacteria
- Status of individual's intestinal microbiome
  - Healthy bacteria provide competition to decrease infectious bacteria from colonizing

Another issue is when one samples fresh food or grocery meats in the refrigerator, bacteria may be detected using PCR or even culture, but at a dose much less than the MID. When this food is taken home and not properly stored at low temperatures, bacteria multiply exponentially and a MID is easily achieved. Cold storage and food processing reduces growth of potentially harmful bacteria. In contrast, if meat, egg, or even a contaminated vegetable is properly cooked, the bacteria will be killed. This is the concept of milk pasteurization or radiation treatment of food which kills any bacteria that may be present in the raw product before the sale. Also, it can partially explain why food safety issues with plant products such as lettuce or onions often occur since salads are not cooked prior to consumption. Food handling becomes especially important in microbial food safety.

This phenomenon brings up an interesting food safety principle related to batch mixing of contaminated food, such as occurs with milk tanker trucks and in ground beef processing. For a chemical toxin, mixing causes dilution, which reduces the concentration of chemical present. Chemicals also degrade over time. For a microbial contaminant, dilution and mixing causes inoculation of the originally microbe-free food, that over time reaches levels that might approach the MID.

Relating MID to the concepts of zero discussed for chemicals is difficult and often reduces to simply detectability, since subsequent handling of the food can result in growth and true contamination. The MID is a moving target and dependent on the numerous variables discussed above. The general trend to avoid getting sick from microbe tainted food is to minimize the potential for contamination by microbes at every stage of food production as shown in Fig. 10.1. This is especially important at the factory and in shipment for packaged mass-produced food. It also becomes important to avoid cross contamination at the grocery store or restaurant or home food preparatory areas. Dilution and mixing are a toxicologist's friend and a microbiologist's foe.

Recent Food Safety Concerns Involving Interpretation of Zero and Chemophobia

The reader should have a reasonable understanding of how chemical thresholds in food products are established, as well as the difference between a zero-food safety risk versus zero concentration of the chemical detected. The imprecise and even sloppy use of these definitions, coupled to the psychological drive toward the ideal of consuming all natural and pure foods with zero contaminants, and thus zero risk of adverse effects, leads to confusion and communication errors. I call this the zero-fusion meme

In the public eye, safety and analytical zero often merge into simply no detection, which mistakenly assigns adverse risk to a concentration that is safe.FDA has been making continuous efforts to address consumer concerns of chemicals in their food by explaining the methodology behind their assessments. As discussed earlier, a new initiative is to reduce contaminants in food to be "closer to zero", acknowledging that safe exposure levels can be defined and may not be equal to analytical zero. They suggest that when consumers consider the issue of chemical contaminants in food, they seek the following: *More complete information from a credible medical and scientific source that would explain how much of the chemical is in the food, and how much of a food someone would have to eat or drink at a level that is harmful to people.*

This fear of chemicals in food, and the excessive drive to achieve zero chemical exposure and risk, has been referred to as *chemophobia,* defined as *an aversion to, prejudice against or even irrational fear of chemicals and chemistry*. It is a psychological aversion or prejudice and is not a true clinical psychiatric phobia. This author would consider it a cultural trend that is reinforced in the media, and is not seriously challenged by the consumer reading a news article or advertisement claim, because these claims generally offer an indirect *confirmation bias* supporting the "problem with chemicals"; in other words, a safe marketing scheme that blames chemicals. This belief originates from a reasonable concern over the potential adverse effects of synthetic chemicals, a feeling consistent with the precautionary principle previously discussed. This is amplified by the irrational fear that these chemicals might cause cancer, a disease discussed in the last chapter that elicits anxiety paired with a drive to avoid any level of exposure to any chemical, although we now know cancer is a multistage disease. In food safety, the potential dangers of chemical contamination get further amplified by the true dangers from microbial food contamination. A similar phobia exists for genetically modified food and nanomaterials, though no true risk has ever been demonstrated.

Chemical-free is immediately associated with zero, which now translates to absence of the chemical and not necessarily absence of an effect. This detection of a chemical suggests impurity. This chemical free cultural trend easily blends with the zero meme, that adds to the mystique this strange number possesses in all fields encountered.

Consumer product labels including pure, natural, or chemical-free offer validation of this fear because they imply products with synthetic chemicals must be harmful. The irony of this fear of chemicals is that as discussed in the chemistry chapter, all life is composed of chemicals. Beneficial life-saving pharmaceuticals are chemicals. Additionally, many chemicals are intentionally added to food, though their names are associated with positive thoughts of health promoting vitamins, minerals, and dietary supplements. As extensively discussed in my earlier book on Chemical Food Safety, some of this fear originates from many chemical contaminants being only identified by their full chemical names rather than common names that may have positive associations. One example will suffice. Vitamin A is commonly named retinol, whose chemical name is (2E,4E,6E,8E)-3,7-dimethyl-9-(2,6,6-trimethylcyclohex-1-en-1-yl)nona-2,4,6,8-tetraen-1-ol. What would the typical consumer think if that was written on the label of your daily bottle of vitamins with no common name provided? At approved doses, Vitamin A is an essential vitamin but at excessive doses, it causes liver disease as well as neurological and skin conditions. Again, the dose makes the poison.

Many chemical contaminants do not have the luxury of having simple common names. Most EPA approved pesticides are innocuous at exposure levels found in food or used around the home, yet they may be dangerous when exposed to during the manufacturing process or when handled or dispersed by agricultural workers loading tank trucks or airplane crop dusters. In public surveys of chemical residues found in food products, trace levels of pesticides strike fear in consumers. Yet these are the same pesticides that are commonly sprayed around the house to kill ants, roaches, or other insects. The residues being reported in food are usually below legal thresholds, at levels millions to billions of times lower than encountered in their routine household product use.

Throughout evolution, our ancestors consumed poisons from plants in their diets allowing the development of protective mechanisms including metabolism to reduce or eliminate potential adverse effects. These processes have been honed over evolution and have led to defensive responses such as hormesis. Such systems do not discriminate between natural and synthetic chemicals. Our diets have always been laced with natural chemicals. When plants were learned to be too toxic for consumption, they were employed as poisons as discussed in the beginning of the toxicology chapter and thus avoided in food. Most agricultural chemicals are derivatives of organic molecules found in natural sources, this field of chemistry being called pharmacognosy.

The average diet contains over a gram and a half of natural pesticides. Many of these natural pesticides, integral components of most of our natural plant-based foods, have been determined to be rodent carcinogens. The problem is that they can't be removed from food as many are integral components of a plant that prevent other adverse health conditions from occurring (*e.g.,* pest infestation, fungal growth, *etc.*) to the plant. In contrast, daily exposure to all synthetic pesticides in the average diet is only about 0.1 mg: 10,000–15,000 times lower than the natural pesticide exposure! An early estimate of cancer risk from natural *versus* synthetic pesticides suggested that risk from natural pesticide exposure was 1000 times greater than from synthetic pesticides. [*See quote on natural compounds by Dr. Bruce Ames at beginning of the previous chapter*]. This number has surely changed due to the decreased use of synthetic pesticides and increased adoption of all natural organic diets.

The apple is my favorite example of a food loaded with both nutrients and toxins. Apples are one of many fruits whose seeds contain acute toxic doses of cyanide compounds. Apples, like cabbage as well as coffee, also contain the potential rodent carcinogen caffeic acid (IARC Group 2B carcinogen). Pitted fruits, such as apricots and peaches, and even sweet almonds contain the cyanogenic glycoside, amygdalin, the primary component of the fad anticancer drug, Laetrile. It can be poisonous to children who ingest it since cyanide is its primary metabolite. When was the last time you picked up an apple or apricot that contained a warning label indicating a potential for toxin exposure? An interesting side is that although caffeic acid may be a potential carcinogen, human exposure to it from its various sources has not detected any evidence for adverse effects in humans.

In my previous book, which was focused on food safety, since the advent of synthetic pesticides in the mid-twentieth century, life expectancy has increased amongst the general population. The earlier and more environmentally persistent pesticides that had been associated with more toxicity have largely been removed from active use. Overall, total pesticide usage is continuing to decrease. Thus, if our massive exposure to synthetic chemicals in our food supply were truly toxic, the opposite trend would have been seen. This would add credence to the efficacy of the risk assessment practices that was discussed in this chapter. *Regulating zero safety risk rather than zero chemical detection seems valid.*

The USDA NRP surveys meat, poultry, and egg producers to assess contamination of drugs, pesticides, toxins such as aflatoxin, arsenic, cadmium, and lead as well as known environmental contaminants such as dioxin and PCBs. The FDA inspects milk products, while the USDA and EPA conduct a pesticide data program to assess pesticide levels in food produced from crops. A list of drugs, pesticides, and other chemicals that the NRP monitors is listed in Table 10.2. It must be noted that many of these chemicals are not approved for use in the US but may appear in imported foods or were once approved but now are not. The NRP also utilizes multi-residue analytical techniques that allow for multiple chemicals to be assayed in the same sample.

Table 10.2 Drugs, pesticides, chemicals and metals monitored by the National Residue Program in 2020[a]

Abamectin	Danofloxacin	Leucocrystal violet	Pyraclostrobin
Acephate	DCCD	Leucomalachite green	Pyrethrin
Acepromazine	DDD	Levamisole	Pyridaben
Acetamiprid	DDE	Lincomycin	Pyriproxyfen
Alachlor	DDT	Lindane (BHC gamma)	Quinoxaline carboxylic acid
Albendazole	Deethylatrazine	Linuron	Ractopamine
Aldicarb	Diazinon	Malachite green	Resmethrin
Aldrin	Dichlorvost	Malathion	Ronidazole
Aluminum	Diclofenic	Manganese	Salbutamol
Amikacin	Dieldrin	Meloxicam	Salinomycin
Aminohydantoin	Difenoconazole	Metalaxyl	Sarafloxacin
Amoxicillin	Difloxacin	Methamidophos	Selamectin
AMOZ	Diflubenzuron	Methomyl	Selenium
Ampicillin	Dihydrostreptomycin	Methoxyfenozide	Semicarbazide
Apramycin	Dimethoate	Metolachlor	Simazine

(continued)

Table 10.2 (continued)

Arsenic	Dimetridazole	Metribuzin	Spectinomycin
Atrazine	Dipyrone	Metronidazole	Streptomycin
Azaperone	Diuron	MGK-264	Strontium
Azinphos methyl	Doramectin	Molybdenum	Sulfachloropyridazine
Azoxystrobin	Doxycycline	Monensin	Sulfadimethoxine
Barium	Emamectin	Morantel tartrate	Sulfadoxine
Benoxacor	Endosulfan	Moxidectin	Sulfaethoxypyridazine
Bifenthrin	Enrofloxacin	Myclobutanil	Sulfamerazine
Boron	Eprinomectin	Nafcillin	Sulfamethazine
Boscalid	Erythromycin A	Naphthol	Sulfamethizole
Buprofezin	Ethion	Narasin	Salfamethoxazole
Butorphanol	Ethofumesate	Neomycin	Sulfamethoxypyridazine
Cadmium	Fenbendazole	Nickel	Sulfanilamide
Carazolol	Fenoxaprop	Nonachlor	Sulfanitran
Carbadox	Fenpropathrin	Norfloxacin	Sulfapyridine
Carbaryl	Fipronil	Norflurazon	Sulfaquinoxaline
Carbofuran	Florfenicol	Omethoate	Sulfathiazole
Carboxylic acid	Flubendazole	Orbifloxacin	Sulfisoxazole
Carfentrazone	Flunixin	Oxacillin	Sulprofos
Cefazolin	Fluridone	Oxazolidinone	Taleranol
Cephapirin	Fluroxypyr	Oxychlordane	Tebufenozide
Chloramphenicol	Gamithromycin	Oxyphenylbutazone	Tefluthrin
Chlordane	Gentamycin	Oxytetracycline	Tetrachlorvinphos
Chloroneb	Haloperidol	Penicillin G	Tetraconazole
Chlorothalonil	Heptachlor	Pentachloroaniline	Tetracycline
Chlorpromazine	Hexachlorobenzene	Pentachlorobenzene	Thallium
Chlorpropham	Hexazinone	Permethrin	Thiabendazole
Chlorpyrifos	Hexythiazox	Phenylbutazone	Thiamethoxam
Chlortetracycline	Hydroxycarbofuran	Piperonyl butoxide	Thiobencarb
Chromium	Hygromycin B	Pirimiphos	Tildipirosin
Cimaterol	Imazalil	Pirlimycin	Tilmicosin
Ciprofloxacin	Imidacloprid	Prallethrin	Tolfenamic acid
Clenbuterol	Indoxacarb	Prednisone	Trifloxystrobin
Clindamycin	Ipronidazole	Profenofos	Triflupromazine
Clothianidin	Iron	Promethazine	Tulathromycin
Cloxacillin	Ivermectin	Pronamide	Tylosin
Cobalt	Kanamycin	Propachlor	Tylvalosin
Copper	Ketamine	Propanil	Vanadium
Coumaphos	Ketoprofen	Propetamphos	Virginiamycin
Fluvalinate	Lascolid A	Propiconazole	Xylazine
Crystal violet	Lead	Propionylpromazine	Zilpaterol

[a]Chemicals listed are parent compounds and not necessarily the marker residue

The vast majority of samples across all food commodity groups have undetectable or below actionable tolerance levels. Yet, consumer "food basket" surveys identify pesticides using the incredible sensitive analytical tools but at levels below tolerances, sounding the greater than zero detection alarm. Yet recall, that *tolerances are a few thousand times lower than the NOAEL.* The NOAEL *is* the only measured metric that has been shown to have no adverse effects seen after lifetime consumption of this amount of chemical in food. It's safe!

Ironically, while wrapping up the writing of this book, I saw the following headline across multiple media platforms: "Chemical found in Cheerios, Quaker Oats, and other oat-based foods linked to potential health issues". This headline was based on the results of a small pilot study by Temkin et al. The culprit was the presence of chlormequat residues in oat products around the LOD, as well as the detection of very low levels in the urine of about 100 adults sampled. Chlormequat is a plant growth regulator that thickens and shortens the height of targeted plants. It is used on food crops such as oats, barley, and wheat. Its LD_{50} in rats is almost 700 mg/kg, and some developmental toxicity has been detected in laboratory animal studies at the *highest* doses studied. The EPA tolerance ranges from 10 to 40 ppm (10,000–40,000 ppb). The levels detected in the above study ranged from non-detectable to about 300 ppb, a value over 300 times less than the established tolerance, which we know from discussions above are already hundreds to thousand times lower than the NOAEL. This compound is approved for use by EPA, CODEX as well as the EU at slightly lower MRLs (2–0.5 ppm), a value still orders of magnitude above the concentrations detected in crops. There is no argument that the chemical is present in both oats and people, at what one would consider non-zero levels, but at LOD—LOQ levels. This seems a timely example of how detecting low concentrations of a pesticide in food, well below tolerance levels, should be regarded as a low safety risk. Instead, it is flagged because the pesticide was simply detected, a hook attractive to the press. Numerous other examples of this meme in action can be easily found throughout the media.

Popular Literature on the Hazards of Eating Toxic Food As I was researching material for this book, I was struck by a popular book whose title was *A Consumer's Guide to Toxic Food Additives: How to Avoid,* renamed from an earlier edition called "*Badditives*". In the opening of the book, the discussion focusses on serious microbial food contamination issues, like *Salmonella,* and the general poor nutritional value of many of the additives and preservatives that make up the modern American diet. However, both of these examples are not *chemical toxins* that we have come to realize, when named as such, are immediately associated with true poisons activating the chemophobia meme again.

In "*Unsafe at Any Meal*", the focus is on food additives, components and preservatives in food as well as solid based nutrition. Yet in a discussion of acute pesticide toxicity to humans, the documented example of toxicity is not food intake at residue levels but rather exposure to pesticides sprayed in the home! Another study discussed purported pesticide intake must have come from food despite any data on pesticide levels in food. A few years ago, I was involved in a case of children dying from the inadvertent use of undiluted commercial pesticide concentrates in a house. Similar incidents have been reported in some Mexican resort hotels. We are aware that a high dose exposure to pesticides is toxic to humans, but not at the sub-tolerance residue levels found in foods.

A final foray into the popular literature, *"Non-Toxic: Guide to Living Healthy in a Chemical World,"* gives advice on how to avoid all synthetic chemicals in your life. Although Paracelsus is cited, it is argued that he is incorrect on the statement that only the dose makes the poison, as there are exceptions. No mention is made that these are not exceptions but are predicted by Haber's law; dose and length of exposure are critical. Some chemicals are simply potent or are allergens that have their own dose response relationship. The adverse effects discussed are not toxic, as we would associate as being poisonous in a classic toxicological sense but are rather food components that are nutritionally unbalanced or unsound. As an example, in a lead paragraph of Chap. 4 of this book subtitled Chemicals in our Food, one reads:

> Scientists have discovered that ingredients such as sodium, high fructose corn syrup, and trans fats, as well as synthetic food additives and preservatives, increase the risk of obesity, diabetes, hypertension, stroke, heart and liver disease (collectively referred to as metabolic syndrome) and immune system disease in both adult and children. Other chemicals in food and food packaging can contribute to endocrine disorders, infertility,—.

The one point that is consistent across this popular literature base is that our modern world is flooded with chemicals, and our diet is often far from ideal. Our western diet and lifestyle, especially the American one, is characterized by an increase incidence of chronic disease but is also characterized by the largest fraction of centenarians in human history. We live longer, but not necessarily as healthily, as in one's youth. Is this due to chemicals or just natural aging?

This author wants to make a distinction between low level chemical exposure *versus* the presence of components, additives, and supplements in processed food at much higher concentrations. For example, salt is an essential nutrient but unhealthy at high levels consumed daily. The diseases associated with such diets include obesity, diabetes, heart and fatty liver disease among others. They are metabolically based, due to consumption of nutritionally unbalanced diets, and associated with continuous consumption of such food, as well as often being associated with correlated life-style behaviors such as smoking, alcohol consumption, and sedentary behaviorThese conditions and their causes are very different than the low *near zero* exposure to chemicals contaminants in food that is the focus of this chapter.

The identification of what specific component of processed food, that is some specific additive or processing agent, excess fat, sugar, and salt, or other factors that are responsible for these chronic diseases, is under intensive study. Potential *"bad-ditives"* that have been purported to cause health effects include:

- aluminum
- lead
- mercury
- artificial sweeteners aspartame and saccharin
- the preservatives BAH and BHT
- food coloring dyes

- the thickener carrageenan
- high fructose syrups
- the so-called meat-glue transglutaminase
- tenderizer monosodium glutamate (MSG)
- partially hydrogenated oils—trans fats

The biggest scientific conundrum is simply that exposure to all synthetic and natural chemicals is complex. It is next to impossible to identify any single culprit, but in reality, there probably isn't a single bad actor. An additive may promote overconsumption of unbalanced foods with the resultant health effects seen in the American diet. A bad diet selection automatically removes consuming what could be a healthy diet. These are truly nutritional issues regarding assessing a healthy balanced diet *versus* a heavily processed junk-food diet, that everyone can agree is unhealthy. Eating a diet solely composed of healthy kale is unhealthy. Although calling them *toxic* or *unsafe* captures media attention, it just feeds into the chemophobia meme. But consuming such chemicals in food for a few days or weeks will not have a serious health impact. For our discussion on what zero means and residue levels of chemicals, the unhealthiness of these diets is immaterial for pesticide residues at sub-tolerance levels, that will be found in organic and so-called balanced diets as well. They are not toxicological hazards.

Recall that zero may also appear on a complex food label indicating that no compound has been added, an example being zero sugar diet soda. In this case, the content is not analysed for the presence of sugar, it simply is not an added ingredient in the recipe.

Some True Chemical Hazards Although the focus of this book is on defining zero, I feel it important to acknowledge that there have been food contamination issues that resulted in violative (greater than safety thresholds) levels in various food commodities. Some of these are naturally occurring, including arsenic, the fungal products aflatoxin and ochratoxin, and the plant toxin aristolochic acid, responsible for Balkan nephropathy. Others are due to synthetic chemical contamination, such as melamine in baby formula discussed below. Marine predatory fish, such as tuna, can naturally accumulate mercury in their tissues. All of these have been detected by regulatory agencies as being violative and requiring action. Allergens are always of concern. Allergic drugs, such as penicillin, are monitored, safe levels defined, and their labeled use takes this into consideration. In contrast, natural allergens, such as peanuts, are not assayed in the product but dealt with by using labels indicating their potential presence. The current crises in the press include microplastics and PFAS chemicals, both essentially showing up everywhere, including food, water and human tissues. Exposure is known but the hazard at the exposure levels measured, and not at high acutely toxic concentrations so far studied in laboratory animals, has not been clearly identified.

Melamine was an economic adulterant. This nitrogen containing plastic was added to infant milk formula in China to deceive nutritional protein assays into reading higher levels of protein than were present due to the nitrogen in the melamine masquerading as protein. Lead has been detected in municipal water supplies at violative levels. It also has been found to be at 200 times the allowable limit in some infant food products (*e.g.*, applesauce) which have components (cinnamon) that could accumulate lead naturally from the soil. Alternatively, lead may have been added to cinnamon to increase the weight of the spice to simply increase supplier profits. Lead has also been added to turmeric to make it yellower. This is economic adulteration, produces levels above tolerances, and must be closely guarded against. This is a serious issue.

The concern with many of these exposures compared to the trace level chemical contaminants that was focused on earlier, is that exposure may be continuous as in water or contaminating the only food source available as with any baby food or especially infant formula. Infants are far more susceptible to toxin effects, especially during development. As this book was written, the FDA is increasing surveillance of imported food due to lack of monitoring in some exporting countries. The impetus for this surveillance was detection of violative residues in import samples, and hence an increase in safety recalls.

A news flash also occurred while writing this chapter about the volatile organic compound ethylene oxide, a known carcinogen, in air around production facilities in the SE Louisiana industrial corridor. Most samples in air were above 10.9 ppt which exceeds the EPA acceptable cancer risk of 100 in a million excess cancer risk. Some samples were above 100 ppt, which translates to a 1000 in a million excess cancer risk. In contrast to the food example with chlormequat where all samples were well below the established tolerances, this finding of *greater than tolerance levels* of a known carcinogen in the air, with potentially long periods of exposure, should be of concern.

Conclusion

We have reached the end of our discussion of what zero means from both a toxicological and analytical chemistry perspective.

- If no chemical is found using the most sensitive analytical techniques available, that is lower than LOD, there is *zero* chemical present and thus no risk of adverse effects.
- If a chemical is found at any concentration below the Tolerance or MRL, there is *zero* risk of toxicity, and this would get scored as *zero* violative residues.

The problem that arises is if a chemical is detected at any concentration, independent of what its safety threshold is, the media and other outlets activate the zero fusion meme that cries danger because a trace amount of chemical was detected.

What is the solution to this flood of chemical exposure in the modern world? Eat a diverse and balanced diet obtained from different sources, an approach that will ensure that one is not constantly exposed to a contaminated food supply. Separate true chemical toxicity (*e.g.*, organ failure, cancer, *etc.*) from simply unhealthy nutritional additives, or an unbalanced fatty or carbohydrate diet. Be aware that if you go to solely organic or vegetarian diets, the likelihood of exposure to both natural and synthetic pesticides increase.

Selected Readings

Baynes RE, and Riviere JE. *Strategies for Reducing Drug and Chemical Residues in Food Animals: International Approaches to Residue Avoidance, Management and Testing*. New York, John Wiley, 2014.

Bonvie L, and Bonvie B. *A Consumer's Guide to Toxic Food Additives*. New York, Skyhorse Publishing, 2020.

Cohen A, and vom Saal FS. *Non-Toxic: Guide to Living in a Chemical World*. New York, Oxford University Press, 2020.

Dabrowski WM, and Sikorski ZE. *Toxins in Foods*. Boca Raton, CRC, 2004.

Dufault RJ. Unsafe at Any Meal. Garden City Park, Square One Publishers, 2017.

Entine J. *Scared to Death: How Chemophobia Threatens Public Health*. New York, American Council on Science and Health, 2011.

FDA. *Is Food Safe if it has Chemicals*? fda.gov/consumers/consumer-updates/food-safe-if-it-has-chemicals Accessed June 7, 2024.

FSIS,—NRP. *Residue Sampling Plan: Fiscal Year 2020 Blue Book*. https://www.fsis.usda.gov/node/1982#:~:text=One%20such%20performance%20element%20is,matrix%20(meat%20or%20poultry). Accessed June 12, 2024.

Jusko WJ. Use of pharmacokinetic data below lower limit of quantitation values. *Pharmaceutical Research*. 29: 2628-2631, 2012.

Klassen C. *Casarett & Doull's Toxicology: The Basic Science of Poisons*, 9th Edition. New York, McGraw Hill, 2018.

Kroes R, and Kozianowski G. Threshold of toxicological concern (TTC). *Toxicology Letters*. 127: 43-46, 2002.

Lin Z, Vahl CI, and Riviere JE. Human food safety implications of variation in food animal drug metabolism. *Scientific Reports*. 6:27907, 2016.

National Research Council, *Scientific Criteria to Ensure Safe Food*. Washington, National Academy Press, 2003.

Pussa T. *Principles of Food Toxicology*. 2nd Ed. Boca Raton, CRC Press, 2014.

Riviere JE. *Comparative Pharmacokinetics: Principles, Techniques and Applications,2nd Ed*, Ames, Wiley-Blackwell, 2011.

Riviere JE. *Chemical Food Safety*. Ames, Iowa State University Press, 2002.

Robinson ES, Tehrani MW, Yassine A, Agarwal S, Nault BA, Gigot C, Chiger AA, Lupolt SN, Daube C, Avery AM, and Claflin MS. Ethylene Oxide in Southeastern Louisiana's Petrochemical Corridor: High Spatial Resolution Mobile Monitoring during HAP-MAP. *Environmental Science & Technology*. 58: 11084-11095, 2024.

Winter R. *A Consumers Dictionary of Food Additives*. New York, Three Rivers Press, 2009.

Zad N, Tell LA, Ramachandran RA, Xu X, Riviere JE, Baynes RE, Lin Z, Maunsell F, Davis J, and Jaberi-Douraki M. Development of machine learning algorithms to estimate maximum residue limits for veterinary medicine. *Food Chemical Toxicology,* 179: 113920, 2023.

Chapter 11
Doping and Performance Drugs, Substances of Abuse, Death, Finance and Other Encounters with Zero

> Please, compete in the spirit of fair play, mutual understanding and respect. And above all, please compete cleanly by refusing doping. (Jacques Rogge)

The goal of this book is to survey the role of zero and nothing across human activities. We have come a long way. The previous applied science chapters focused on how analytical chemistry is used for the detection and quantification of substances that could impact human health. This chapter employs many of these same analytical techniques and endpoints but discusses their use in the legal monitoring of chemicals and drugs used in illicit activities. These are not the same endpoints. Also, the use of zero in diverse endeavors ranging from finance to climate change will be briefly overviewed.

Doping

The final technical arena where analytical *zero* plays an important role is in the detection of illegal drugs or other substances in competitive sports as well as in the legal control of substances of abuse (Fig. 11.1). Numerous types of drugs and substances may be taken to increase both human and equine athletic performance, in terms of improved cardiac function, enhanced muscle mass, increased ability to oxygenate blood, or to decrease pain associated with prolonged exercise, all with the goal of giving the athlete a competitive edge or skewing the betting odds in horseracing.

Figure Use Attributions
Figs. 11.1, 11.2, 11.3, and 11.4: Wikimedia – No restrictions
All other figures are either self-drawn, photographed or from author's own publications (attributed)

J. E. Riviere, *Zero – Much to Do About Nothing?*,
https://doi.org/10.1007/978-3-031-82998-7_11

Fig. 11.1 Performance doping in sports

Doping has been around since the ancient Olympic games in 668 BCE, where interestingly, figs were commonly used to enhance a runner's performance based on their nutritional ingredients. Ancient people have also purportedly eaten raw animal testicles to enhance athletic prowess. As pharmacology developed in the nineteenth century, specifically synthesized stimulants and steroids, such as testosterone, were taken along with caffeine and alcoholic beverages. As anabolic steroids became widely available in the 1950s, their legal use was widespread and even referred to by the popular press as the "*Breakfast of Champions*". Doping of horses, originally using arsenic as a stimulant, was first reported in the mid-sixteenth century and continued to be employed in the early nineteenth century. The first well documented use of performance enhancing drugs in horse racing was the 1941 case of the 18-year-old thoroughbred named Holloway, who won multiple races suggesting that performance was improved after testosterone administration.

Allowing or forbidding the use of drugs and nutritional substances as doping agents in competitive sports is a societal, ethical, and legal decision. Substances were routinely and legally used as performance enhancers as little as a few decades ago. In some minds, there is a fine line between acceptable nutritional and metabolic supplementation *versus* illegal stimulants, a scenario that parallels defining the difference between dietary supplements and an FDA approved therapeutic drug. Modern regulations are developed to even the playing field in athletic competitions with some substances being defined as illegal by consensus, this selection being a very contentious process.

The highest profile area that commands public and media attention are in the Olympic Games that began in ancient Greece and in horse races such as the American Kentucky Derby first ran in 1875. It is in these arenas that regulations and definitions are most developed and defined and serve as the legal foundation for other doping regulations. The first modern doping case occurred in the 1904 summer Olympics where athlete Tom Hicks consumed a mixture of cognac and strychnine. Official Olympic regulation of doping started in 1967. To get an idea of the nature of such rules, the 2024 Paris Olympic state:

> It is the athletes' personal duty to ensure that no prohibited substance enters their bodies. Athletes are responsible for any prohibited substance, or its metabolites or markers found

> to be present in their samples. Accordingly, it is not necessary that intent, fault, negligence or knowing use on the athlete's part be demonstrated in order to establish an antidoping rule violation.

This rule has many caveats and technical nuances associated with it. Any level of detection indicating the presence of a banned substance in a sample such as urine or blood, results in a violation unless there is an established *Decision Limit* (*e.g.*, tolerance) for that substance defined in the current governing technical document. This limit is based on the current state of analytical chemistry. For doping violations, the laboratory must *identify* the prohibited substance, its metabolites, or its markers at a concentration greater than the decision limit. Two samples are collected, one for screening and the second for confirmation in the event of a positive test. This is a *hard-zero concentration* indicating absence of the drug.

Analyses must be conducted in laboratories accredited by the World Anti-Doping Agency (WADA) using acceptable international standards. Analytical techniques are identical to those described in the chemistry chapter and now consists primarily of GC and LC MS multiple residue screening assays. In contrast, biomarkers of doping are those altered biological variables that indicate a specific biological response associated with a banned substance has occurred.

The WADA conducts various validation assessments across participating laboratories to ensure accuracy and reproducibility of its testing. Prohibited substances are those listed on the official list of prohibited substances, as well as their metabolites or markers, which are periodically officially updated. These substances are primarily performance enhancing drugs or natural compounds. If the athlete requires a prohibited substance for medicinal use, it can be specifically identified and approved for a *Therapeutic Use Exemption* for that athlete for that event. Contaminants, which may often be found in nutritional supplements, can result in a positive test, but athletes are liable for this occurrence if they are detected. Examples of prohibited drugs and methods include substances that satisfy one of three criteria:

1. They enhance athletic sport performance
2. They represent a health risk
3. They violate the spirit of sport

These include anabolic agents such as steroids (*e.g.*, testosterone), certain bioactive peptides, hormones and related compounds (*e.g.*, erythropoietin, growth hormone), β2-adrenergic agonists (*e.g.*, terbutaline), diuretics, narcotics, cannabinoids, as well as masking agents and strategies including autologous blood transfusions and other oxygen carrying enhancing modifications. If a compound's athletic enhancing effects are only seen after acute exposure (*e.g.*, amphetamines, opioids), the compounds may be prohibited only in-competition for a specific race. Some are prohibited in specific Olympic sports, such as β-antagonists in events requiring fine-motor control such as archery. One should note that scientific evidence of efficacy for most performance enhancing drugs is lacking; however, their ongoing use suggests a belief in their utility and thus a violation of the spirit of good sport.

A very similar set of anti-doping regulations also applies to equine and greyhound racing. Doping in performance horses is defined as the *illegal application of any substance, except a normal diet, that might modify the natural abilities of the horse at the time of a race*. Equine regulations, far from internationally harmonized, have been promulgated by agencies, including the *Federation Equestre Internationale (FEI)* in the form of the *FEI Equine Anti-Doping and Controlled Medication Regulations* and the *Veterinary Regulations, the US Equestrian Federation Equine Drugs and Medications Program,* as well as the recent Federal Trade Commission's *Anti-Doping and Medication Control Program* administered through the *Horseracing Integrity and Safety Authority (HISA)* and enforced by the *Horseracing Integrity Welfare Unit (HIWU)*. In many cases state regulations take precedence over federal regulations. Substances are divided into *controlled medications* that are allowed outside of race day and *banned substances* that are never allowed to be used in a horse intended to race.

To facilitate the legal therapeutic use of a controlled drug that may be banned during an actual race, a delay termed the *detection time* can be recommended. This defines the time after drug administration when it should not be detectable in a urine sample for a typical horse at a concentration greater than the *screening limit* established for that compound. This concept is very similar to that described in the last chapter for establishing withdrawal times in food animals, except the matrix is now urine rather than milk or edible tissues. The endpoint is now actually LOD rather than tolerance or MRL. This is a practical guideline attempting to aid the veterinarian to establish a withdrawal time for the horse's trainer to follow.

Detection times are only guesstimates that incorporate safety factors to ensure that any therapeutic drug that was administrated is not detectable on race day. They are not legal metrics because *detection of a banned substance above its LOD, even after following a delay after dosing, will immediately disqualify the horse from racing*. If only detected post-race, the winning title may be rescinded. This occurred in the 2021 Kentucky Derby where the original winner Medina Spirit was stripped of the title after testing positive for the corticosteroid betamethasone at a concentration of only 21 pg/ml.

Similar rules and practices apply in numerous human and animal sport competitions, including but not limited to track and field events, swimming, weightlifting, baseball, cycling, and greyhound racing. Large scandals have occurred in major league baseball with the use of anabolic steroids and human growth hormone. Such violations blocked many athletes from entering the baseball Hall of Fame, for example the detection of steroid use by Barry Bonds, Roger Clemens, and Mark McGwire. In all doping cases, solid science and legal precedence govern activities and decisions. In human competition, it is the individual athlete responsible for a violation. Due to the extremely high sensitivity of doping assays, inadvertent exposure to banned substances may occur requiring the athlete to screen food and their environment for potential substances. In the animal world, it is the team of trainer, veterinarian, and owner responsible to ensure the equine athlete remains drug free.

Unfortunately, doping is a major part of equine racing, due to its deep roots in gambling and sport betting. This will become more prominent with the continued proliferation of real-time on-line sport betting that makes even obscure races potential targets of illegal fixing. Unlike in human sports, substances may be given to a horse to throw a race by causing a specific horse to lose. Because of the impacts of doping on race outcome or on medal winnings and national prestige in human competition, the practice of doping is highly regulated.

Zero and Detectability Nuances

The literature and history of sports doping is a fascinating subject but far beyond the focus of our discussion for this book.

> Unlike the thresholds established for food safety and toxicology endpoints, assaying in doping is conducted for a fundamentally different purpose: did the athlete take banned performance enhancing compounds? The endpoint is detection rather than quantification. The goal is zero concentration and proof of the absence of drug in the sample.

As can easily be appreciated, the increase in analytical capabilities makes detection much more sensitive. Although this can ensure that banned substances are not taken, it can also result in detection of substances far below any potential biological effect in the human or animal athlete. Examples can include ingestion of poppy seeds in food that triggers an opioid violation or ingestion of pork tainted with β-agonists, anabolic steroids, or glucocorticoids. In these cases, the concentrations of drug in the athlete are far below those required for athletic enhancement but are easily detected as non-zero by a doping screening assay.

The attempts to hide the use of banned drugs in athletes are monumental, often involving innovative approaches to chemical manipulations and masking their detection in urine. These techniques are vigorously pursued by athletes, gamblers, and in some cases even nation-states in the pursuit of national prowess. Equally pursued is their detection by regulatory authorities charged with making the sports competition fair for all participants. A current case is the banning of Russian athletes in the 2024 Olympics because of prior doping violations. Methods to mask doping include molecular manipulations of a drug to retain biological enhancing activity but avoid detectability in screening assays. Urine is a preferred matrix to screen for banned substances because the kidney is the prime excretory pathway for many drugs and their metabolites (See Fig. 7.1 and associated discussion). This results in urine concentrations being much greater compared to those of blood because they are concentrated in the process of excretion.

Another issue that has important implications are false positives as well as false negatives. This occurs when, for example, the analytical detector being used cross reacts with another substance closely related to the target compound generating a false positive result. These errors occur more often with multi-compound screening assays. This is combatted by running a second confirmatory assay usually using a different analytical method when a positive is detected in the screening assay. However, if a false negative occurs, a second assay will not be performed because there is no signal that additional testing is needed. This could occur when a compound binds to the container or some other substance in the urine. There are numerous mechanisms for such errors to occur, but quality control protocols of modern analytical assays generally reduce their occurrence. Random testing of athletes also minimizes this outcome.

The pursuit of zero is the true endpoint of doping assays because detection and not just quantification is pursued. There is often no physiological enhancement or effect seen, detection the drug test itself is the endpoint. The goal is to be as sure as possible that an athlete did not take a banned substance. Detection of a defined prohibited substance or its metabolite is the sole goal. This results in the definition of zero being dependent on the analytical capabilities of the authorities responsible for policing the sport.

Drugs of Abuse

Another application of drug testing that has both serious legal implications and nicely illustrates the conceptual nuances of zero, is the screening of urine samples for drugs of abuse including opioids and cannabinoids such as marijuana. Such tests are often conducted as part of employee screening protocols. Alcohol screening of one's breath is performed on drivers suspected of being intoxicated. In this latter case, sufficient historical data has been available using a breathalyzer to statistically define two endpoints of alcohol abuse.

- *Legal limit of intoxication* (*e.g.*, threshold) of 0.08% that defines that the adult driver is inebriated.
- *Legal limit for zero tolerance* ranges from 0.0% to 0.02%, which is applied to underage drivers because consumption of alcohol by minors is illegal.

The zero-limit applied to underage drivers is meant to reflect that no measurable amount of drug was in the body. This is like doping in athletes. Detection equals violation. For adults who can legally consume alcohol, the test is essentially one for detecting inebriation and intoxication at a level that makes driving dangerous. This is a concentration linked to a functional endpoint. As with most other biological scenarios, there are functional and analytical definitions of *zero* or *no effect* that are context dependent and need to be precisely defined for their proper application.

End of Life

The final biological encounter with zero that will be discussed is at the end of life, where organ function ceases. We have metaphors of zero in terms ranging from *flat-lining* in cardiac arrest to having *zero brain function* in the condition known as *brain death*. These are very complex issues with severe measurement limitations. As discussed in the biology chapter, the body is made up of trillions of cells and scores of organs, all of which may live independent existences for various intervals of time before the entire organism truly dies. One can appreciate this when organs are preserved by individuals participating in organ transplant programs where brain function has been judged to be lost. *This brain-centric paradigm defines life as being associated with possessing or potentially regaining a functioning cognitive brain reflective of a mind* (Fig. 11.2).

Brain death is generally defined as the complete, permanent and irreversible loss of brain functions that control cognitive (primarily cerebral cortex) as well as involuntary autonomic system activity (primarily brain-stem) that are needed to sustain life in the rest of the body.

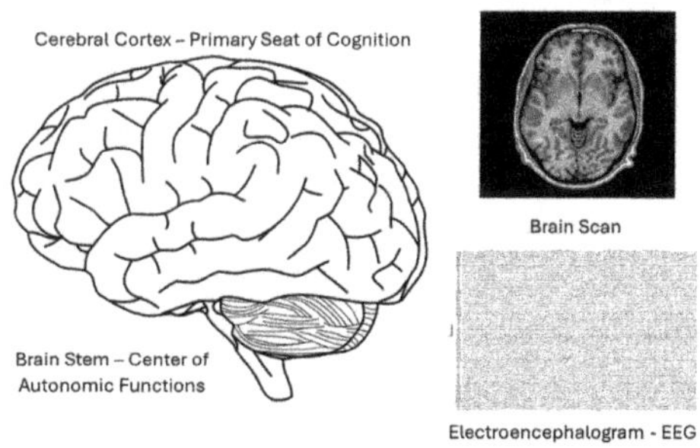

Fig. 11.2 Brain regions and tools to assess its function

This is different from a *persistent vegetative state* where autonomic functions persist, but central cognitive brain function has ceased. It also differs from non-terminal states where both subconscious cognitive and autonomic brain functions continue under certain conditions such as a *coma* and *locked-in-syndrome*. The assessment of functional brain activity is multimodal, ranging from bedside physical examinations and tests, to electroencephalograms that can measure activity in different brain regions, to various types of modern functional brain imaging scans. Recent advances reported as this book was being completed (Bodien at el., 2024) included studies showing one in four patients unresponsive to commands showed neural activity measured by electroencephalograms and functional magnetic resonance imaging, a condition known as cognitive motor dissociation.

The major factor impacting decisions of resuscitation is whether any level of residual cognitive brain functions remain when the body is on artificial exterior life support systems. These definitions are fluid because of the complexity of individual regions of the brain, along with varied medicolegal definitions in different jurisdictions written for different purposes, in addition to the continued advancement of brain science and neurology. International organ transplant guidelines have driven laws that define legal death, many of which were based on a 1968 Harvard Medical School report defining irreversible comas. These have evolved over time with advancements in medicine.

Cardiac arrest, which is characterized by a flat electrocardiogram (ECG), reflects the loss of heart function that if it persists, causes hypoxia in all organs and ultimately brain death. This can be assessed by approaches ranging from a lack of a digital pulse, no heart beat upon auscultation to a flat ECG. For our purposes, the longest one can survive with a nonfunctioning circulatory system incapable of pumping blood throughout the body is in the range from 5 to 10 min.

Brain death would be defined as *zero electrical activity* measured by an EEG while cardiac arrest would be zero electrical activity in the heart on an ECG. There is a difference in what we mean by zero electrical activity in a human *versus* the inability to detect electricity in a copper or silicone-based electrical circuit or battery where sensitivity of a detector may be important. In a biological organism, electrical activity is generated as a result of intact cells and organ function, producing what is called an action potential. The strength of this electrical impulse is relatively constant for different organ functions (*e.g.*, a nerve impulse versus cardiac pacemaker cell, *etc.*). They do not show a linear decay to zero as seen in an electrical system when a battery fails. Instead, they are more of a step function where when alive, a specific impulse of a defined current will be generated, but when dead, nothing will be generated. This zero truly means zero heart or brain function. Loss of electrical activity in the brain suggests there is nothing functional left in the mind.

Finance

Zero also finds itself in countless economic scenarios, including *zero interest rates, zero net worth* or a *zero-profit* condition. These definitions are usually straightforward and for many scenarios do not suffer from instrumental measurement errors as did many of the previous examples in science and technology. An individual's *net worth or wealth*, or a business' *book value* or *shareholder's equity* is calculated as *assets minus liabilities*. When tabulated solely in cash holdings, the calculation is straightforward. However, many assets do not have an easily estimable value as they are often dependent upon market conditions at the time of sale. Others are based on expected earnings or are intangible assets that are associated with a business's reputation. All of these must be estimated and thus are prone to estimation error. Such considerations become important with the sale of a property or in events such as bankruptcy.

In investments, a *zero rate of return* could be calculated as when the income received only balances costs of production. A zero rate of return on an investment would be zero income produced or zero net income after inflation is considered. *Zero percent financing* means that one only pays back the amount of money lent with no interest payments due. These are commonly promotional interest rates to persuade consumers to buy a product such as a car or appliance. A *zero coupon* or discount bond is one where interest payments are not made over the life of the bond because the bond is bought at a steep discount to its face or par value. The difference between the purchase price and par value is the rate of return. In stock trading, a *zero-day option* (#0dte) allows an investor to bet on whether a specific stock will rise or fall before the close of the market that day.

A *zero-profit condition* is the condition that occurs when an industry or type of business has an extremely low (near-zero) cost of entry to or exit from the industry. This reflects a state of perfect competition. Zero profit also refers to a business where income is balanced by explicit and implicit costs to the point that all resources are being efficiently used.

Zero-based budgeting is an accounting method that starts each new budget period at a zero baseline by not considering past spending. The goal is to allocate income to specific costs and purposes so that at the end of the period, there is zero unspent money. These purposes could include a defined salary or profit for a business owner, an individual's savings goals for emergency or retirement needs, or planned replacement of capital. The approach assures that one's expenses perfectly balance off one's income. In the world of academic research grants and contracts, this is the budgeting strategy required where income is actually the original award amount.

This sampling of economic and finance definitions is obviously only a cursory survey of a vast quantitative discipline where zero is encountered in numerous calculations. *The key to interpreting zero in these economic and financial contexts depends on having an operational definition to define the scenarios where it can be applied, and the accuracy required to estimate zero* (Fig. 11.3).

Fig. 11.3 The scourge of bankruptcy

The financial zero that has the greatest impact is when a person's or business' net worth approaches zero, and results in *bankruptcy*. Zero becomes a threshold where the failure to be able to pay bills or the costs of living or operations require immediate action to be taken. Having zero net worth resulting in bankruptcy can have serious consequences. Life changes! In the early nineteenth century, people were sent to debtor prison. Today, property and other assets may have to be liquidated. For an individual, this condition impacts credit scores, which then affects the ability to obtain credit cards, loans, mortgages, and even employment. If a loan is obtainable, the interest rates on borrowed money will be greater. For a corporation, its very existence might be in jeopardy or a reorganization with cancellation of existing debt gives it a fresh start. Other numbers might spell trouble, but again zero is a true threshold that has serious consequences and supports the adversity and nervousness that is associated with it.

Climate Change

A final context where zero has been encountered in modern society is that of *zero emissions* associated with the attempt to reduce pollution levels in the atmosphere that would stop global warming. Global warming has been identified as one of the existential threats facing modern society (Fig. 11.4). In addition to the effects that oppressive heat has on animal and plant life, 90% of excess heat is absorbed by the oceans which potentially contributes to extreme weather events including droughts, storms, tornados and hurricanes.

Fig. 11.4 The many faces of global warming

The component of global warming attributable to human activity can be controlled by reducing the quantities of so-called *greenhouse gases* released into the atmosphere which traps heat and warms the planet. These gases include carbon dioxide, methane, nitrous oxide, and water vapor with carbon dioxide making up many of these emissions. It is believed that at the *net zero point*, where the rate of addition of greenhouse gases into the atmosphere (largely by the burning of fossil fuels) is balanced by their removal through carbon sequestration processes (such as absorption by forests, plants, and the ocean), global warming would stop. Of course, there are negatives to this process including carbon dioxide induced acidification of the ocean. The goal of reducing input could be achieved using renewable energy sources such as nuclear, solar, and wind. *Net zero emissions* is the balancing point between release and absorption of carbon into the atmosphere, more akin to the concept of a zero in a mass balance equation, an object's center of gravity, a financial balancing of assets and liabilities or zero's location in the number line between negative and positive numbers (Fig. 1.1). It is the approach to maintaining a climate equilibrium at current levels.

In contrast to this *net zero* endpoint, aiming for a *real zero* endpoint would require avoiding the emission of all climate-damaging gases as well as ending all other practices that contribute to global warming and resource depletion. The difference is that *net zero*, as the name implies, cancels out emissions with sequestration processes while *real zero* eliminates emissions entirely.

Conclusions

This chapter wraps up the discussion of how zero is applied across multiple spheres of life. As discussed throughout the text, this unique number has the tendency to focus attention upon itself in a manner very different than other numbers. It is a threshold between something and nothing, between an effect and no effect; an illicit drug was taken, one is bankrupt, the earth is warming, or your loved one has passed. Although these examples all seem to trend toward negative outcomes, a zero in a drug test is a positive event for both the athlete and sport involved. In finance, zero is a positive when debt doesn't exist or when one snags a zero-interest loan. As was stressed throughout this book, context is crucial to interpreting zero.

Selected Readings

Aikin R, Baume N, Equey T, and Rabin O. Biomarkers of doping: uses, discovery and validation. *Bioanalysis*. 12:791-800, 2020.

Bodien YG, Allanson J, Cardone P, Bonhomme A, Carmona J, Chatelle C, Chennu S, and Schiff ND. Cognitive motor dissociation in disorders of consciousness. *New England Journal of Medicine*. 391: 598-608, 2024.

Bowers LD. The analytical chemistry of drug monitoring in athletes. *Annul Review of Analytical Chemistry*. 2:485-507, 2009.

Brealey R, Myers S, Edmans A, and Allen F. *Principles of Corporate Finance*, 14th Ed., New York, McGraw Hill, 2023.

Fragkaki AG, Kioukia-Fougia N, Kiousi P, Kioussi M, and Tsivou M. Challenges in detecting substances for equine anti-doping. *Drug Testing and Analysis*. 9:1291-1303, 2017.

Helm D. *Net Zero: How We Stop Causing Climate Change*. Glasgow, William Collins, 2021.

Holt RIG, Erotokritou-Mulligan I, and Sönksen, PH. The history of doping and growth hormone abuse in sport. *Growth Hormone & IGF Research*, 19: 320-326, 2009.

International Olympic Committee. *Anti-doping Rules applicable to the Games of the XXXIII Olympiad,* Paris, 2024.

Johnson M. *Spitting in the Soup: Inside the Dirty Game of Doping in Sport*. Boulder, Velo Press, 2016.

King LA. *Forensic Chemistry of Substance Misuse: A Guide to Drug Control.* London, Royal Society of Chemistry, 2009.

Melicher RW, and Norton EA. *Introduction to Finance: Markets, Investments, and Financial Management*. 16th ed., New York, Wiley, 2016.

Michael A. History of brain death as death: 1968 to the present. *Journal of Critical Care*. 29:673-678, 2014.

Thevis M, and Schanzer W. Examples of Doping Control Analysis by Liquid Chromatography-Tandem Mass Spectrometry: Ephedrines, β-Receptor Blocking Agents, Diuretics, Sympathomimetics, and Cross-Linked Hemoglobins. *Journal of Chromatographic Science.* 43: 22-31, 2005.

Toutain PL. Why the Racing Industry and Equestrian Disciplines Need to Implement Population Pharmacokinetics: To Learn, Explain, Summarize, Harmonize, and Individualize. *Drug Testing and Analysis,* April 29, 2024.

Truog RD, Paquette ET, and Tasker RC. Understanding brain death. *Journal of the American Medical Association*. 323: 2139-2140, 2020

US Equine Federation. *Guidelines and Rules for Drugs and Medications*, 2022.

Watson CJ, Stone GL, Overbeek DL, Chiba T, and Burns MM. Performance-enhancing drugs and the Olympics. *Journal of Internal Medicine*. 291:181-196, 2022.

Wu HH, Liu WY, and Huang MC. *Moving Toward Net Zero Carbon Society*. London, Springer Nature, 2023.

Chapter 12
Summary and Personal Conclusions

> The point about zero is that we do not need to use it in the operations of daily life. No one goes out to buy zero fish. It is in a way the most civilized of all the cardinals, and its use is only forced on us by the needs of cultivated modes of thought. (Alfred North Whitehead)

I started this journey into zero since I have wrestled with its meaning my entire professional career. It is a number that is mathematically unique and even peculiar at times. It psychologically harbors connotations and emotions that are simply not associated with other numbers. My principal reason behind writing this book was to try and better understand where this psychological angst arose and further probe the meaning of zero and nothing in biology and especially food safety. My concept of zero did not seem to match that reported by the media and on social platforms. Why has the meaning of zero become so complicated?

Historically, the emergence of zero enabled the development of modern mathematics. In numerous fields, when a calculation yields an answer such as 1, 4, 19 or 587, the brain is not startled, no sleep is lost, and the result is simply recorded. But if a calculation yields a result of 0, attention becomes focused, and the calculation is often rechecked. This is especially true when that result enters into another equation where a 0 might be lethal. In some circumstances, a 0 result causes extreme anxiety, if it is attached to one's bank account or investment gain. Alternatively, when 0 is the result of a chemical assay for a drug in food or a urine drug test, one is reassured of its positive outcome.

One could argue that the number three is also as special and interesting as zero. Just consider the Catholic Holy Trinity. The digit three's uniqueness is also due to it being the number of vertices in a triangle, a fundamental shape in nature, as well as the number of dimensions we live our daily lives in. The field of *Trigonometry* only deals with triangles. As Mark Parker writes in his book on triangles which make the case for three being a unique number; *Triangles are everywhere, and everything is triangle.* And of course (*e*) is important as the base of natural logarithms. Finally, there are also books on infinity (∞) and pi (π), though this author still holds that the number 0 commands unique attention across multiple disciplines.

Figure Use Attributions
Figs. 12.1, 12.2 and 12.3: Wikimedia – No restrictions
All other figures are either self-drawn, photographed or from author's own publications (attributed)

J. E. Riviere, *Zero – Much to Do About Nothing?*,
https://doi.org/10.1007/978-3-031-82998-7_12

When zero is viewed as its position on the number line, additional uses for it in everyday life become evident. When one counts down to zero, something happens. An obvious use of a zero with an impact is in the countdown to launch a rocket (Fig. 12.1). In this case the countdown starts from the left or negative side of the number line up to zero:

Fig. 12.1 Rocket launches—Falcon 9 and Saturn V

T-10, T-9, T-8, T-7, T-6, T-5, T-4, T-3, T-2, T-1, and O: Lift-off!

Positive times then relate to describing events after takeoff and in flight. This is an obviously important number and in this case zero does not reflect that nothing is happening!

Another familiar case where zero is encountered is the elevator. One would think this is straightforward, but the zero floor could be either the ground floor or the basement. In fact, it can get very confusing as North America, Europe and other regions adopt different approaches. In America the ground floor is generally the first floor and the basement zero or simply B. In Europe and some other countries and architectural schemes, the ground floor is zero or G, the first floor is now above ground and labeled as one, while the basement or B is below. This ground floor reflects the easiest exit from a building. Sub-basement floors are named using various approaches including negative numbers, letters or simple names such as basement or lobby. This can be appreciated by the elevator and floor buttons in Fig. 12.2. There really is no consistency, but in this case, zero does not imply nothing, in fact far from it, as it implies the base and foundation of subsequent floors. One could argue that in elevators, the floor that truly is nothing is the 13th floor which most elevators also skip purely because of superstition (triskaidekaphobia), the floor in elevator-speak that truly represents nothing!

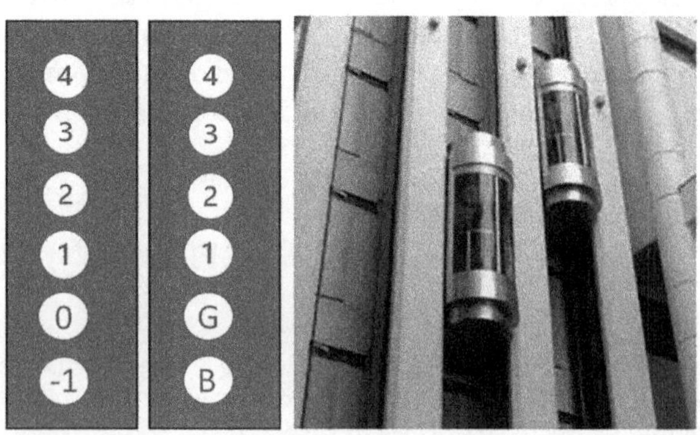

Fig. 12.2 Elevators

In both cases, zero is an important divider: ground floor or blastoff. It is unique and establishes the framework and origin of other numbers, be it floors or time in flight. In any countdown, zero is the target or endpoint, as in boarding an airplane, starting a play, or launching a rocket. An important number indeed, and in these cases totally devoid of an association with nothing or empty.

As discussed throughout the book, interpreting the meaning of zero and nothing, thus depends on the context of the question posed. When one contemplates the meaning of nothing in philosophy, numerous implications arise. What is the difference between empty and nothing? Empty implies the existence of a vessel or container that contains nothing in it. That container could even be the physical extension of the entire universe. If one is counting, empty means zero objects, a clear result. But nothing could also mean there is no vessel present to begin with. As we saw in our discussion of Eastern philosophy when examining Rubin's vase (Fig. 2.2), the question arises as to what is real: the two faces or the vase? Similarly, is a window the frame or the empty transparent space it encompasses?

I feel it important to briefly summarize from a personal perspective some of the concepts of nothing and zero that were objectively presented in detail throughout the book. The introduction broadly surveyed where zero is found, but since now the nuances in its interpretation have been discussed, I feel it appropriate that some of my own conclusions and opinions be presented. *The most important point I learned from this exploration was that context is a crucial determinant of meaning, requiring operational definitions to be known before any conclusions are drawn.*

Philosophy and History

We started this intellectual voyage from antiquity to the present by tracing the emergence of both the concepts of nothing and the void, as well as the emergence of the mathematical number zero. The first obvious finding was the difference in acceptance of the concepts of nothingness and the void in Western *versus* Eastern philosophies and religions. This distinction delayed the emergence of the number zero in Western science to well into the second millennium. How much of a role did Christianity play in suppressing the concept of nothing and zero to not enter science earlier? Although speculative, did Western religion's acceptance of the philosophies of Aristotle and Plato, rather than those of Democritus and Epicurus, strongly influence non-acceptance of the number representing the void? Aristotle's maxim that "nature abhors a vacuum" as well as his numbering system, which began with the digit 2, held great sway in the nature of the cosmologies and mathematical systems that could emerge throughout this cloistered period. Did this lack of a concept of nothing or empty in Western language inhibit the concept of a zero from ever being imagined in terms of the concepts inherent to linguistic relativism or Wittgenstein's postulate that language limits one's thoughts?

In contrast, Eastern philosophies always incorporated the concept of the void and nothing into their core beliefs. It is little wonder that zero first emerged in the Indian subcontinent centuries before the concept was accepted in the West. Similarly,

Eastern philosophical concepts of inaction as the baseline for personal behavior parallel both the modern statistical principle of the null hypotheses as well as the regulatory philosophy underlying the precautionary principle. The philosophical concept of nothingness persists in most Eastern philosophies as well as in modern schools of existentialism.

Although this very early history of zero has no direct effect on modern concepts of mathematics, it continues to hold a psychological effect when zero is thought about or encountered, examples being:

- Is there going to be a problem with further calculations?
- Will this crash my calculator?
- How will I pay the rent?
- I just lost a small fortune in roulette because of that darn set of zeros.
- Phew, I passed the drug test!
- Time has run out.
- Food is safe to eat as no residues were detected.
- It's zero degrees outside.
- We have zero-tolerance for that behavior.
- Blast-off.

As science continued to develop, this *zero or nothingness meme* affected the meaning attached to it in novel situations that were vastly different than those previously encountered. In the twenty-first century, we still grapple with the meaning of zero in fields ranging from physics, toxicology, medicine, to sport regulations. Zero is a common tool used in advertising (recall Fig. 1.1 in Introduction) to convey concepts ranging from zero interest to zero calories to ground zero to zero contaminants, suggesting purity and perfection. Its use is unavoidable.

Mathematics and Computational Pitfalls

We discussed unique properties of zero as a number as well as its distinct functions in mathematics. Its origins rest firmly in its use as a placeholder that allowed for the development of our modern decimal numbering system. The point 0,0 is the intersection point of the Cartesian X,Y coordinate system, which is foundational to modern math.

Yet for all these positives, zero also can be a menace in calculations, as outlined in Table 4.1. One is always on the lookout for zero appearing in a denominator or as the operand of a logarithmic expression. However, zero is also a savior in simplifying complex equations since if it appears in a multiplicative term, the entire equation turns to zero. Its emergence allowed for the development of both algebra and calculus. The use of limits to calculate what happens when a term approaches zero; allowed for vast swaths of modern science and technology to develop.

Because of these issues with the appearance of zero in equations, care must be taken when using instruments that report zero as a result, a situation common in analytical chemistry and when weighing objects. This is especially true when such data are automatically uploaded to a database. The widespread use of automated databases and cloud storage raises issues when zero values are scraped from the web and removed from their original context. This will be more pronounced with the increased use of language-based generative artificial intelligence programs.

Physics and Nature

The question that has puzzled philosophers and scientists for centuries is simply what does nothing really mean in the physical world? As discussed earlier, the concept of the void panicked some, yet inspired others. When modern physics looks at the question of whether space can truly be empty, one must define the pertinent physical scale, and whether matter or energy is being considered. Empty is particularly tied to the scale at which it is measured (Fig. 5.4). At the quantum scale, energy fields dominate that may produce particles from quantum fluctuations. It is here that concepts, such as Planck length and Planck time, are descriptive and define the granularity of space. It's where absolute zero becomes contemplatable. At the larger atomic, molecular, and microscopic scales; particles dominate. Here, the physical definition of empty, nothing, or the void can be measured and are only constrained by instrumental limits. A laboratory vacuum contains a million more particles than the environs of deep space. Yet, both are not truly empty and devoid of all particles. Similarly, when zero light is detected, one is cast into darkness. Darkness for humans is not darkness for instrumented humans or many of our animal relatives. *For many of the points and phenomenon discussed in this book, it is the mesoscopic scale starting at a few hundred nm up to the macroscopic scale of μm to mm where chemistry and biology exist.* Properties at this scale are often emergent from those below, yet laws that apply at lower scales are not directly applicable to phenomena that occur at a higher scale.

One explanation for why zero continuously needs attention is that it is constantly redefined by advances in technology across all fields. This ranges from the measurement of absolute zero in physics to the increased sensitivity of analytical instruments in chemistry to the growth of novel financial assets which require new definitions for net worth or financial risk management. In all science, instrumental sensitivities define the lower limits of the scientific definitions of zero. They define the inference space or context where an experimental determination of zero can be applied. Because of never-ending technical advances, modern instruments continually increase the sensitivity and precision of all our measurements. This forces one to be cautious when using new data in older models or algorithms that may not have been defined at the current level of sensitivity as they may have different inference spaces. As discussed in earlier chapters, this is a major issue with older regulations applied to modern methods of detection.

Physics also deals with the concept and measurement of time. In some contexts, time is simply the 4th dimension of space. It is astronomically defined based on the Earth's orbit around the sun, which defines months and years or the rotation of the earth that defines the time of day. The invention of mechanical timepieces leads to the psychological impression that time passes in quanta and slowly ticks away. Here zero time implies an instantaneous event or interval. Finally, cosmology deals with the singularity and time zero or the Royal zero. In many cosmologies, zero becomes both the beginning and the end.

Chemistry, Biology and Life

Biology and chemistry operate in the mesoscopic scale bridging the molecular and microscopic biologically-relevant domains. It is in this domain that the tension exists between detection of a chemical and whether that chemical can exert a biological effect. As discussed extensively in these chapters, exposure to a chemical or detection of a chemical does not automatically equate to triggering a biological or toxicological effect. Remember, dose and duration make the poison.

The body is a complex organism at all scales examined, ranging from sub-cellular architecture to cells to organs to the entire body. There are complex pathways involved in absorbing a chemical from the environment or in the food, being distributed throughout the body, and then actually getting to the site where an adverse effect may occur. These are driven by diffusion, face dilution throughout the body, and possibly enzymatic destruction. This results in all biological effects being dose related with most having quantitative dose to activity relationships expressed by the sigmoidal dose response curve (Fig. 7.9). The overall assessment then of what specific dose is required to have an effect is determined by numerous types of bioassay systems ranging from cell cultures to populations of animals. *The zone of biological effect is bracketed by the highest dose reflecting lethality, the* LD_{50}*, to the lowest dose producing no biological effect, the NOAEL.*

The issue arises when safe levels in food, such as tolerances and MRL, are set. The take-off point from the actual experimental data to start theoretical back-extrapolation is the NOAEL. The NOAEL is further reduced by tenfold safety or uncertainty factors to result in the allowable daily intake, the ADI. This is then used to determine what *lifetime* exposure would result in no harm. This safe concentration sets the maximum acceptable exposure level (Fig. 8.7) which in individual food types is the tolerance or MRL. If a food source exceeds these tolerance levels, then harmful residues may be present. *But this would only be potentially harmful if consumed for life and this number is still at least one hundred-fold less than the actual concentration that was experimentally determined to produce no adverse effects after chronic consumption.* The chance that a single food source would remain contaminated for decades and that an individual would eat this same food for decades is vanishingly small. A very conservative food safety bias is engrained into this system. A good strategy to avoid any lifetime exposure to any contaminant is simply to vary sources of food. However, if a water supply is contaminated, then long term exposure could become a serious issue.

The tension arises when a chemical such as a pesticide is found in food at levels below the acceptable tolerance but still at a quantifiable amount (Fig. 6.8). There is no quantifiable biological risk to exposure at this concentration. Yes, it was detected as being present. There are hundreds of other natural compounds and toxins that may be present at higher levels that are not even looked for and do not produce an adverse effect. It's even worse when a very low concentration is detected (LOD) but below the quantifiable level (LOQ). Some media report this as "dangerous chemicals were detected in the food." Many of these agricultural chemicals, such as pesticides and plant growth promotants, are used to improve crop production, ward off plant diseases or kill mycotoxin-producing fungus. They may be toxic to humans but at the much higher doses only encountered in manufacturing or dispensing, not at residue levels. *Detection simply does not translate to risk of adverse effects.*

At low concentrations near LOQ and LOD, estimation errors also increase and add to the uncertainty. Difficulties remain in interpreting values at LOD or below because although not quantifiable, they do indicate the presence of a compound. They are essentially zero, but zero often can't be entered into mathematical equations that use exponential decay to calculate risk factors. This remains a computational and statistical quagmire. Issues also occur when older regulations define exposure at the LOQ that was defined when the regulation was originally written. This is true for carcinogens. Many of these are decades old and the modern LOQ may be a thousand times below the original value. The solution is to define terms, such as De Minimus level or Threshold of Toxicological Concern, to allow low level of compounds to be scientifically accepted.

The endpoint is different when drug assays are used to regulate sports doping. Here *detection proves exposure and thus potentially illegal drug use.* There is no safety endpoint. In detection of alcohol in people, there is alcohol detection in a minor *versus* quantification in an adult, which only prohibits levels related to intoxication, because adults are allowed to consume alcohol at levels determined not to affect driving ability. *Drug testing is a detection issue.*

In today's complex agricultural food production networks, the use of synthetic chemicals is widespread. Chemicals are everywhere but all that are used in animal and plant production have been screened for safety. Concern is now focused on forever chemicals like PFAS and micro-plastics in both the environment and our bodies. Are they harmful at the levels detected? Natural toxins are everywhere, and chemicals are voluntarily taken in nutritional supplements and so-called functional foods. Highly processed foods containing numerous chemicals can be hazardous to health, are really a nutritional issue rather than a toxicological one. Short term consumption will not be lethal.

- Is there a trade-off chasing zero in all scenarios *versus* accepting some low level of exposure and focus resources in other areas?
- What are the economic benefits of too much *versus* too little testing?

Accidents do happen and can result in chemical contamination at levels above safety thresholds that are of concern. These should not be discounted but they are not at low residue levels. Similarly, some industrial chemicals may accumulate and although not of concern two decades ago, are of concern now. Infrastructure ages, that may also result in contamination of water supplies such as seen with lead pipes. *These are toxicological issues that deserve attention and action.*

A large reason for public concern about chemical exposure in food and the environment is one of communication. *There is not a universal definition of zero*. The context in which it was determined and in which it could be applied must be defined. Grocery basket surveys often only report detection independent of what the allowable limits may be. When a concentration of a chemical is found, writers have difficulty suggesting there is zero biological risk. Internet searches determine that the chemical detected may be harmful, but no one looks at the dose that any adverse effect was seen at, *versus* the level of contamination detected. The public must be made aware of this dilemma as the situation will only worsen with increasingly sensitive analytical methods. The chemicals haven't changed. The focus should be on issues with a higher positive public health impact. Unfortunately, in today's political climate, these issues are not openly and freely debated.

Summing Up

We have covered zero from many perspectives and in many fields, each with their own context and baggage. Unfortunately, there is often an unintentional crossover between disparate fields. Definitions are vague and their inference spaces are not well defined and never articulated. The various memes we discussed associated with zero and nothing often get mixed together, resulting in formation of mixed metaphors and blends of idioms. In such scenarios, nonsensical conclusions are made. I found this numerous times in researching this book using the tools of generative AI. It's amazing how often zero and nothing from diverse sources become entangled!

The breadth of fields where zero and nothing are encountered is astonishing. My hope for readers of this book is that when zero is encountered in whatever venue, they consider the context of how zero was generated compared to the context of the situation where it will be applied. And when someone remarks that nothing is there, think a little about what that really means.

The quote from the Dalia Lama on the front piece of this book elegantly and succinctly states the dilemma that defines zero: *"A zero itself is nothing, but without a zero you cannot count anything; therefore, a zero is something, yet zero."*

I will end this book with an important life skill inspired by this encounter with zero and nothingness. *Niksen*, the Dutch art of doing nothing, is an important life skill said to promote happiness, regain one's composure, find calm, and prevent burnout. It can be a short break that fosters relaxation and allows one to regain focus. It can be simple daydreaming. In Italy, it is *dolce far niente*—the sweetness of doing nothing. At this point as I finish this book on nothingness, I believe I will take a nap and enjoy some niksen! (Fig. 12.3)

> The horizon of many people is a circle with a radius of zero. They call this their point of view. (Albert Einstein)

The End

Fig. 12.3 Daydreaming man by Edmond Lebel

Selected Readings

Mecking O. *Niksen: Embracing the Dutch Art of Doing Nothing*. New York, Harvest, 2021.
Parker M. *Love Triangle—How Trigonometry Shapes the World.* New York, Penguin, 2024.

Index

J. E. Riviere, *Zero – Much to Do About Nothing?*,
https://doi.org/10.1007/978-3-031-82998-7

www.ingramcontent.com/pod-product-compliance
Lightning Source LLC
Chambersburg PA
CBHW061639030826
49219CB00003B/39

9783031829970